NEW DIRECTIONS FOR PROGRAM EVALUATION
A Publication of the American Evaluation Association

15-10-91

Nick L. Smith, *Syracuse University*
EDITOR-IN-CHIEF

Evaluation and Privatization: Cases in Waste Management

John G. Heilman
Auburn University

EDITOR

Number 51, Fall 1991

JOSSEY-BASS INC., PUBLISHERS, San Francisco
MAXWELL MACMILLAN INTERNATIONAL PUBLISHING GROUP
New York • Oxford • Singapore • Sydney • Toronto

H
62
A 1
V 51
1991

EVALUATION AND PRIVATIZATION: CASES IN WASTE MANAGEMENT
John G. Heilman (ed.)
New Directions for Program Evaluation, no. 51
Nick L. Smith, Editor-in-Chief

Microfilm copies of issues and articles are available in 16mm and 35mm, as well as microfiche in 105mm, through University Microfilms Inc., 300 North Zeeb Road, Ann Arbor, Michigan 48106.

LC 85-644749   ISSN 0164-7989   ISBN 1-55542-777-4

NEW DIRECTIONS FOR PROGRAM EVALUATION is part of The Jossey-Bass Education Series and is published quarterly by Jossey-Bass Inc., Publishers (publication number USPS 449-050).

EDITORIAL CORRESPONDENCE should be sent to the Editor-in-Chief, Nick L. Smith, School of Education, Syracuse University, 330 Huntington Hall, Syracuse, New York 13244-2340.

Printed on acid-free paper in the United States of America.

Contents

EDITOR'S NOTES

Privatization of public services constitutes one of the major developments or movements (Moe, 1987) in the field of public policy and administration in the past twenty years. In its many forms, privatization commonly involves an increased role for private-sector organizations in the production and delivery of public services. This expanded role has introduced new issues into the organization and management of service programs. Since these issues can directly affect program performance, the evaluation profession has good reason to explore and engage them. The present volume undertakes that task.

The Nature of Privatization

My goal here is to outline the nature of privatization and identify some of the main issues that run through the growing literature on this subject. I consider why these issues should interest evaluators in a range of fields and introduce the four chapters that constitute the body of the volume. In the final chapter, an afterword, I indicate how the main chapters respond to these issues in the context of waste management and suggest some general conclusions for the theory and practice of evaluation.

Savas (1987, p. 3) defines privatization as "the act of reducing the role of government, or increasing the role of the private sector, in an activity or in the ownership of assets." This "load shedding" can take many forms, ranging from sale of government assets to more limited strategies such as deregulation, use of vouchers, franchises, and contracting for services (for a detailed and useful taxonomy of privatization strategies, see Savas, 1990).

Contracting for services is the most prevalent form of privatization in the United States; thus, it defines a policy strategy that many evaluators are likely to encounter in practice. In contracting, the government decides to provide (make available) a service. Through a contract, the "government authorizes and pays [a] private firm to [produce and deliver the] service" (Savas, 1987, p. 68; for a valuable and influential discussion of the distinction between service provision and production, see Kolderie, 1986). The volume of privatization that takes place through service contracting alone is enormous. Savas (1987, pp. 70–74) summarizes evidence that in the mid-1980s hundreds, if not thousands, of local units of government contracted with private organizations to supply scores of services ranging from ambulance service and building maintenance to youth programs and zoning control.

Service contracting and other privatization strategies raise multiple issues that can affect program operations and outcomes in ways of concern to evaluators. Some of these issues are well established in the literature, whereas others have only recently been addressed. Perhaps the most thoroughly explored issues have to do with efficiency and accountability. Theory developed by Savas (1987) and others promises policymakers and program managers that privatization can achieve political goals of accountability as well as economic goals of efficiency. Simply put, the argument states that the dynamic of the private sector, based on market competition, leads to greater economic efficiency and productivity than we would expect from public-sector organizations. Further, public-sector officials who decide to privatize can draft contracts in a manner that ensures public accountability in service delivery.

Evaluation Issues

The evidence for these propositions is far from conclusive. Private-sector firms do not always produce services more efficiently than do their public-sector counterparts (Weimer and Vining, 1989, pp. 174–178). Furthermore, many thoughtful concerns about political and administrative issues have been raised, including the effects of privatization on accountability (Wise, 1990). In general, then, two questions that evaluators may want to place on their agenda for assessing privatization are (1) Does the privatization strategy increase efficiency and productivity? and (2) Does the privatization strategy make adequate arrangements to preserve public accountability in service production and delivery?

These two questions invite refinement: What do efficiency and productivity mean, and why would we expect them to increase under privatization? What does accountability mean, and what kinds of arrangements can secure it? To some extent, these matters depend on the specific details of individual programs. However, some recent contributions suggest potentially useful, broad answers.

The Economic Nature of Goods and Services. Standard texts in evaluation (Posavac and Carey, 1989) and policy analysis (Patton and Sawicki, 1986; Weimer and Vining, 1989) consider the meaning and measurement of economic efficiency and productivity. The general issue is that of comparing resources committed with services or other outcomes produced. The theory of privatization, as Savas (1987) develops it, suggests why we may expect more favorable ratios from private activity than from public. As already stated, market competition is assumed to force efficiency in private firms. Thus, evaluators may want to ask, Does the privatization strategy used in a particular program provide ready access to market competition? Savas suggests an additional question, arising from the possibility that in privatization the economic nature of the good or service in question may change: Does the privatization strategy change the economic nature of the good in ways that affect its production and delivery?

To clarify what Savas's question means, it is helpful to refer to the typology of goods that he has developed. He focuses attention on two issues. The first is how readily a good can be simultaneously used by multiple persons. For example, only one person can receive individual client counseling from a caseworker at a time, but many people can simultaneously hear a public lecturer. The second is how readily some individuals can be denied access to a good if other individuals have access to it. For instance, public roads are accessible to all drivers, while toll roads can exclude drivers unwilling or unable to pay to use them.

These characteristics, excludability and consumption, form the typology of goods shown in Table 1 (Savas, 1987, p. 56). An example of a private good is a shirt: Only one person can wear a shirt at a time, and shirt producers can use market mechanisms to exclude individuals from using the shirts they produce. An example of a common-pool good is a public commons or an underground aquifer. A toll bridge is an example of a toll good, and national security is a collective good. In the latter case, once security exists, it exists or is accessible for all, and it is jointly enjoyed or consumed.

Many goods represent mixed cases: They do not fall purely and neatly into one of these categories. And, many goods and services can "migrate" along the dimensions of jointness of excludability and consumption. Thus, the nature of a good or service can be the object of policy. For instance, as Savas (1987, p. 42) points out, New York City originally operated Central Park as a toll good (by charging admission at gates), but then it opened the gates, abandoned exclusion, and allowed the park to be "used—and abused—as a collective good."

These distinctions can be useful from the standpoint of evaluation because they suggest how the nature of goods and services can affect arrangements for providing them. That is, privatization of public service may change not only the economic nature of services but also the managerial requirements of whatever is being produced and delivered.

Organizational Design and the Problem of Utilization. Beyond the issue of the economic nature of goods and services, several recent publications address the problem of configuring the organizational arrangements that provide public service. Wise (1990) urges increased attention to configurations that join or mix public and private organizations in the process

Table 1. A Typology of Goods
in Terms of Excludability and Consumption

Consumption	Excludability	
	Easy to Deny Access	Hard to Deny Access
Individual	Private goods	Common-pool goods
Joint	Toll goods	Collective goods

Source: Savas, 1987, p. 56 (copyright © 1987 Chatham House Publishers, Inc.).

of public service. The emphasis here is on interorganizational structures and processes. And, in a recent review of four volumes in the *New Directions for Program Evaluation* series, Schneider (1990, p. 393) focuses attention on the related issue of the "organizational context in which evaluation takes place."

The themes to which Wise and Schneider direct our attention can be useful to program evaluators in that they offer access to the issue of program accountability. In brief, the notion of democratic accountability implies that (1) government officials and agencies are respectful of and responsive to the rights, needs, problems, and complaints of citizens and (2) there are established channels of power and control to promote such responsiveness. Institutional arrangements to support accountability appear at all levels of governance. At the constitutional level, the federal system, separation of powers, and checks and balances all serve this purpose. At the level of service provision, the bureaucratic, political, and organizational issues that Wise and Schneider address can be translated into design issues of accountability.

There is extensive debate in the evaluation field, however, concerning the status of bureaucratic, political, and organizational issues. The debate continues to play out in terms of utilization, but the arguments advanced appear relevant to the substance of evaluative research as well. In brief, Patton (1986, p. 30) suggests that utilization occurs "when there is an immediate, concrete, and observable effect on specific decisions and program activities resulting from evaluation findings." Weiss (1988a, p. 6) suggests that Patton's emphasis on direct, immediate application of results is of limited use for evaluation beyond the framework of "relatively small programs without much divergence of interests." She adds that "most evaluations are undertaken in a policymaking system where authority is dispersed, multiple groups have a say, and policy is the resultant of conflicts and accommodations across a complex and shifting set of players." In this context, the nature of utilization changes dramatically: "Evaluation findings that become known in the larger policy community have a chance to affect the terms of debate, the language in which it is conducted, and the ideas that are considered relevant in its resolution" (1988a, p. 10).

Weiss (1988b, pp. 20–21) also calls attention to how the organizations that might use evaluation results actually work. That is, they tend not to behave as rational decision makers. Rather, they act "according to bureaucratic rules and standard operating procedures, or through organizational politics (with factions or subunits vying for advantage), or through 'garbage-can' processes (where a decision is the almost-chance confluence of the streams of participants, problems, solutions, and choice opportunities that are flowing through the organization at the moment), or acting first and crafting explanations of their actions afterwards, retrospectively labeling them 'decisions.' "

This debate is relevant to the evaluation of privatized programs. Patton (1986, pp. 301–307) refers to and appears to accept the same images of organization that Weiss invokes. But they disagree on the meaning of these images for evaluation. Patton's reaction is to focus on specific and immediate use by identifiable individuals who care about evaluation and have the authority to apply its results. In practice, this application of results seems most likely to occur where authority is relatively well fixed within an organizational setting.

When interorganizational configurations become a central feature of service production, as they certainly are in the case of privatization (Wise, 1990), decisions about using information to change programs seem likely to involve the multiple groups and subtle processes to which Weiss refers. Thus, in the context of privatization, Weiss's conception of utilization becomes especially plausible and applicable.

Moreover, if we sense that organizational arrangements affect the ways in which programs will be adjusted in the light of new information, is it not also reasonable to consider organizational configurations as features of the programs being evaluated? That is, organizational issues can make an important difference not only in the use of evaluation research but also in the program activity that the research investigates.

A growing literature indicates that organizational processes and outcomes, and their management, are not the same in the public sector as in the private sector. Thus, there is special reason for program evaluators to look at organizational issues that come into focus or change as programmatic mass shifts from one sector toward the other. Several recent contributions suggest what to look for as shifts occur. Evaluators who want to read more about these matters should find the following works helpful.

Kettl (1988) directs attention to the political and managerial issues that arise under privatization or "government by proxy." He suggests that successful service delivery depends less on the *economic* characteristics of the good or service involved and more on two *political* and *administrative* features of a program: (1) whether the service providers and takers have a clear agreement on the nature and objective of the service and (2) whether there is an effective institutional framework for ongoing monitoring of service production and delivery. These ideas can be of practical use to evaluators because they point to specific questions about individual program arrangements: Do the "partners" in the public and private sectors have a "meeting of the minds" concerning privatized service production and delivery? Have they made effective arrangements for ongoing monitoring of private-sector operations by the public organization in question?

Yet other considerations come into focus in Wise's (1990) warning against ignoring the fit between organizational arrangements for service delivery, on the one hand, and operation of specific privatization approaches, on the other. Wise also emphasizes that service production and delivery are likely to involve multiple organizations based in different sectors, and that this arrangement carries consequences for the issue of accountability.

From the perspective of program evaluation, Wise (1990) raises several questions concerning the impact of organizational factors on program performance: What organizations contribute, directly or even indirectly, to service delivery? What impact, if any, do the structure and processes of these organizations have on specific program operations? If program activity depends on coordination across organizational boundaries, are the relevant public and private organizations "configured" in a manner that supports service production and delivery? Though stated here in general terms, these questions can be tailored to the organizational details of specific programs. The potential productivity of carefully configured interorganizational arrangements becomes apparent in the work of Kash (1989). He argues that in a few sectors of our economy (agriculture, medicine, and military industry) we have developed imaginative and integrative arrangements powerful enough to support the entire sector through what he calls "synthetic innovation." In this process, public- and private-sector organizations mesh, mix, and evolve into a stable and cooperative framework. While he does not directly address issues of program evaluation, Kash puts on the table a positive question to ask about the organizational foundations of privatization: Does the privatization strategy foster synergy between the different dynamics that energize public and private organizations, respectively?

To summarize, recent contributions to the literature of public administration and policy science extend the privatization debate beyond the important and established issues of program efficiency and accountability. An emerging theme concerns the organizational contexts in which privatization strategies play out. Depending on its design and operation, a given program may change the configuration of public and private organizations through which a service has been produced and delivered. When this happens, the conditions of successful policy implementation and program management may change also.

An Evaluation Opportunity. These issues can productively inform program evaluations. There is little reason to believe that the volume of privatization will dwindle during the 1990s. Public resources to pay for services are tightly constrained, while public demands for services show no sign of lessening. The constraints on federal and state budgets are massive in both political and economic terms, and they are rooted in at least three decades of political evolution in the United States. "In its embrace of privatization, the Reagan doctrine [of limiting federal government to providing services but not producing them] represented not so much the beginning as the culmination of a trend in federal administration and management of domestic programs" (Seidman and Gilmour, 1986, p. 119).

Among the more prominent economic factors driving privatization are the Gramm-Rudman deficit reduction mandates, the savings-and-loan bailout, the costs of financing U.S. military activity in the Middle East, persistent trade deficits (driven recently by the high costs of imported oil), and the continuing public resistance to higher taxes. Beyond these economic matters, ideological motors also drive the privatization movement. The fuel for these motors is a blend of belief in the inherent efficiency of the private sector with belief in a limited role for government (Heilman and Johnson, in press). These factors translate into powerful pressures to move program costs off the federal budget, down to the state and local levels of government, and then out toward the private sector (Johnson and Heilman, 1987).

As the private sector engages the production and delivery of public services through the strategies of privatization, the issues outlined here come into play. The evaluation literature provides relatively little systematic discussion, and relatively few research examples, to assist the evaluation of privatized programs and the policies framing those programs. The present volume is designed to offer such assistance. The discussion so far has addressed the nature of privatization and the issues that are prominent or emergent in the relevant policy literature. Three points bear emphasis: (1) Privatization is a broad movement that includes multiple strategies for producing and delivering numerous types of public service. (2) The origins and growth of the privatization movement suggest that it will continue through the 1990s and that evaluators will have many opportunities to address it at all levels of government. (3) The privatization literature suggests several questions that can plausibly inform evaluative research. Recent contributions have extended the debate on efficiency and accountability to include the organizational arrangements through which privatized programs are implemented and managed.

The substantive essays in this volume, *Evaluation and Privatization: Cases in Waste Management,* address these matters in the single policy area of waste management, as broadly construed. The field of waste management is well suited to this exercise. It is an area in which evaluative research will be needed to inform the policy process. Many prominent problems of waste management raise controversy over public health and the costs of promoting it. Some of these problems include selecting sites for toxic waste dumps, paying for toxic waste cleanups, allocating responsibility across federal, state, and local governments, managing high-level and low-level radioactive waste, disposing of solid wastes as landfills are saturated or become prohibitively expensive, stopping marine damage caused by ocean dumping, cleaning municipal wastewater, assuring the quality of drinking water, and controlling non-point-source pollution, such as that caused by runoff of agricultural chemicals and wastematter.

These and other waste-related problems of public health and safety seem likely to remain prominent during the 1990s for three reasons. First, the problems are not going to go away. Second, public opinion has become increasingly sensitive to these issues. Third, continuing advances in medical technology and science will facilitate the detection of new waste-based health problems for public policy to address. In brief, there is good reason to expect that evaluative research will be needed in, and can contribute to, policy development and program management in this field.

Private-sector actors have taken on a range of roles in the design and implementation of waste-management policies and programs. Engineering firms have built and own merchant (speculative) sewage-treatment facilities; private firms have become active in all phases of solid waste collection and disposal, including landfilling and combustion; private management firms contract to dispose of hospital waste and to operate toxic waste disposal sites.

In sum, problems of waste management are arising in connection with a growing number of public services. Many of these services, such as waste collection, waste disposal, and water treatment, are carried out at the local level of government. Evaluators at this level, and at the state and national levels of government, increasingly must deal with the problems of waste management in some form. And, whatever the specific waste-management problem, private-sector participation in its definition and solution appears likely to be an option.

Overview of the Chapters

The four main chapters in this volume present the results of research on these matters. The chapters vary along four dimensions: the area or type of waste management, the type of privatization or private-sector involvement in the policy process, the type of evaluation issue addressed, and the mix of quantitative and qualitative analysis. In varying degrees, each chapter represents an effort to construct the dynamic play of forces, the subjective reality, with which public- or private-sector players deal. While these chapters have not been written from a self-consciously "naturalistic" (Lincoln and Guba, 1985) or "constructivist" (Guba and Lincoln, 1989) perspective, each reaches in some way beyond quantitative generalization. Together, the chapters help to define an emerging agenda for the evaluation of privatization.

In Chapter One, Laurence J. O'Toole, Jr., addresses a classic question about privatization through direct empirical analysis. Is a public service provided and managed more effectively and efficiently by a public-sector organization or by a private-sector organization? Using his unusually rich data from private industry, O'Toole presents a detailed and theoretically grounded comparative study. In this case, rigorous quantitative analysis is informed by the use of detailed interview data to illuminate the complexity of local contract operations.

The author of Chapter Two, Charles J. Spindler, has conducted exten- sive research in solid waste policy at the international, national, and state levels. The present chapter grows out of his comprehensive evaluation of policy options for the state of Florida. He develops a conceptual framework for classifying government policies and industry initiatives in terms of the extent to which the action on which they depend is voluntary or com- pelled. Taking the case of Florida's Solid Waste Act of 1988, he then explores the dynamics of compelled public-private partnerships. One key element in this process is the manner in which the public-sector policy induces cooperative behavior from private industry. The private-sector initiative is compelled by the conditional nature of the legislation: A man- datory deposit system is imposed *only if* industry fails to meet certain recycling goals. Industry behavior is shaped in part by objective factors that make up the political economy of waste management. Spindler outlines half-a-dozen of the most important issues. He then shows how they can form the basis for a legislative strategy designed to stimulate desired behav- ior by the private sector.

In Chapter Three, John G. Heilman and Gerald W. Johnson report on a subtle but important problem that can arise as private firms take on responsibility for public-service production: In order to preserve public accountability, public agencies need to engage the privatization process and adjust their regulatory policies to the contours of the new public- private configuration. But to what extent can they do so? How do state regulatory agencies engage and respond to privatization? As it turns out, relatively few state agencies have established any formal policy in this regard. Thus, the research described here focuses on informal processes that take place "in the shadow" of codified law and policy. Heilman and Johnson develop the issue, present data based on a national survey of agencies, and then draw conclusions and recommendations.

In Chapter Four, Michael R. Fitzgerald and Amy Snyder McCabe address the problem of selecting high-level, radioactive waste disposal sites under the Nuclear Waste Policy Act of 1982. One major political and tech- nical problem is to ensure the integrity of sites chosen to contain radioac- tive waste indefinitely. The process of evaluation enters this picture through "quality assurance" that the scientific judgments about site characteristics are based on rigorous, reliable, and valid procedures and data. Fitzgerald and McCabe explain in detail why quality assurance faces an especially severe burden of proof under these circumstances. In this politically volatile context, the usual problem of utilization is turned upside down: Rather than receiving too little attention, the results of quality assurance are stretched to the limit in multiple directions by diverse competing interests in the public and private sectors. The story of how quality-assurance results lead to conflict and uncertainty draws together many of the themes already developed in these Editor's Notes.

Finally, in Chapter Five, I indicate how the main chapters of this volume bear on the themes and issues outlined above. I then suggest how these themes and issues may come into play in evaluation across a range of programmatic settings.

John G. Heilman
Editor

References

Guba, E., and Lincoln, Y. *Fourth-Generation Evaluation*. Newbury Park, Calif.: Sage, 1989.

Heilman, J. G., and Johnson, G. W. *The Politics and Economics of Privatization*. Tuscaloosa: University of Alabama Press, in press.

Johnson, G. W., and Heilman, J. G. "Metapolicy Transition and Policy Implementation: New Federalism and Privatization." *Public Administration Review*, 1987, 47 (6), 468–478.

Lincoln, Y., and Guba, E. *Naturalistic Inquiry*. Newbury Park, Calif.: Sage, 1985.

Kash, D: E. *Perpetual Innovation: The New World of Competition*. New York: Basic Books, 1989.

Kettl, D. F. *Government by Proxy: (Mis?)Managing Federal Programs*. Washington, D.C.: CQ Press, 1988.

Kolderie, T. "Two Different Concepts of Privatization." *Public Administration Review*, 1986, 46 (4), 285–291.

Moe, R. C. "Exploring the Limits of Privatization." *Public Administration Review*, 1987, 47 (6), 453–460.

Patton, C. V., and Sawicki, D. S. *Basic Methods of Policy Analysis and Planning*. Englewood Cliffs, N.J.: Prentice-Hall, 1986.

Patton, M. Q. *Utilization-Focused Evaluation*. (2nd ed.) Newbury Park, Calif.: Sage, 1986.

Posavac, E., and Carey, R. *Program Evaluation: Methods and Case Studies*. (3rd ed.) Englewood Cliffs, N.J.: Prentice-Hall, 1989.

Savas, E. S. *Privatization: The Key to Better Government*. Chatham, N.J.: Chatham House, 1987.

Savas, E. S. "A Taxonomy of Privatization Strategies." *Policy Studies Journal*, 1990, 18 (2), 343–355.

Schneider, S. "The Status of Program Evaluation." *Public Administration Review*, 1990, 50 (3), 393–395.

Seidman, H., and Gilmour, R. *Politics, Position, and Power: From the Positive to the Regulatory State*. (4th ed.) New York: Oxford University Press, 1986.

Weimer, D. L., and Vining, A. R. *Policy Analysis: Concepts and Practice*. Englewood Cliffs, N.J.: Prentice-Hall, 1989.

Weiss, C. H. "Evaluation for Decisions: Is Anybody There? Does Anybody Care?" *Evaluation Practice*, 1988a, 9 (1), 5–19.

Weiss, C. H. "If Program Decisions Hinged Only on Information: A Response to Patton." *Evaluation Practice*, 1988b, 9 (3), 15–28.

Wise, C. R. "Public Service Configurations: Public Organization Design in the Post-Privatization Era." *Public Administration Review*, 1990, 50 (2), 141–155.

John G. Heilman is associate professor of political science at Auburn University, Auburn, Alabama. He has served since 1984 as a member of the organizing committee for the National Energy Program Evaluation Conference. Together with Gerald W. Johnson, he has written a forthcoming book on the politics and economics of privatization and is completing a book on state revolving loan funds.

Contracts with private firms offer one mode of alternative provision for public services. Evidence on this approach to the operation of municipal wastewater treatment facilities indicates that performance advantages may be possible for certain types of communities. Savings are possible in larger facilities and when the local government assumes a relatively active and energetic oversight role.

Public and Private Management of Wastewater Treatment: A Comparative Study

Laurence J. O'Toole, Jr.

Currently, governments at all levels face severely constrained resources. Thus, it is no surprise that alternatives to the "standard" mechanisms of public service provision are being intensively explored. One such alternative mode is the practice of contracting for governmental services (Savas, 1987), whereby a public authority, such as a unit of local government, contracts with a private firm to produce or deliver a service. There is much controversy over the alleged benefits of such private management. This chapter considers the costs and performance of contract management of municipal wastewater treatment facilities. While few practicing evaluators may need to address policy or programs directed specifically at wastewater treatment, many are likely to encounter the broader issues that arise in this context.

The chapter begins with a brief review of the subject of contracting. Then some pertinent details of service provision in the field of wastewater

The research on which this chapter is based was financed in part by the U.S. Department of the Interior as authorized by the Water Resources Research Act of 1984 (P.L. 98-242). The contents of this chapter do not necessarily reflect the views and policies of the U.S. Department of the Interior, nor does mention of trade names or commercial products constitute their endorsement by the U.S. government. I acknowledge the invaluable assistance of the company whose contract operations experience forms the basis of the analysis reported here. Local officials in various communities were also very helpful during the course of the field research. I also thank Jean England and Robert Gottesman, graduate research assistants, who ably assisted with coding and analysis of the data.

treatment are summarized. These sections are followed by a discussion of the particular research question and design utilized in this study. Results of the analysis are reported, discussed, and interpreted. Finally, implications for contracting in this policy sector are sketched.

The Service Contracting Phenomenon

Nearly everywhere in the United States, governments now contract with nonprofit agencies and with businesses for at least some services. Contracting for various types of governmental services, especially at the municipal level, constitutes a well-developed practice and is the subject of an established literature. (For relatively early evidence of the importance of the practice, see Florestano and Gordon, 1980; see also U.S. Advisory Commission on Intergovernmental Relations, 1985; Editor's Notes, this volume.)

Many types of contracting arrangements are possible at the local level (Valente and Manchester, 1984). Contracting may involve a relatively simple and limited agreement for some private participation in a public-sector activity, or it may entail comprehensive privatization. (As described by Heilman and Johnson, this volume, comprehensive privatization refers to a wholesale transfer of a public function to the private sector. In the field of wastewater treatment, it occurs when a private firm owns, finances, designs, builds and operates a facility, and bears responsibility for meeting regulatory standards.) The range of contracting options and provisions, including the comprehensive privatization phenomenon, has prompted some efforts at conceptual and analytical clarification (Kolderie, 1986; Savas, 1987; Starr, 1985; Straussman and Farie, 1981). Partially as a result of the conceptual complexity involved, and also because of political disagreements and conflicting empirical findings, the literature on contracting out contains the views of both advocates and detractors. (The former group includes Armington and Ellis, 1984; Savas, 1979. Among the detractors are the American Federation of State, County, and Municipal Employees [AFSCME], 1983; Brudney, 1984; Fitzgerald and Lyons, 1986; Palumbo, 1986; see Brooks, Liebman, and Schelling, 1984, for a representative sampling of diverse views).

Controversy over the alleged efficiency and managerial effectiveness of private firms, and over the broader merits and demerits of private provision of public services, has prompted a number of empirical studies in specific policy fields and across various national settings (for example, Borcherding, Pommerehne, and Schneider, 1982; DeHoog, 1984; Stevens, 1984; McDavid, 1985; Marlowe, 1985; Perry and Babitsky, 1986; Slawsky and DeMarco, 1980; Weimer and Vining, 1989; Wollan, 1986). This previous work suggests that contracting offers advantages to governments in certain policy fields and under certain circumstances. In general, though, overall or blanket prescriptions for (or proscriptions against) contracting are unwarranted. Investigation of the policy features and contextual details of different settings is required if governments are to make sensible decisions about their various options for contracting.

Contracting for Wastewater Treatment Operations

Municipal wastewater treatment constitutes a substantial technical and managerial task requiring very large expenditures of funds. Wastewater treatment works (WTWs) are expensive to build and to operate and maintain. They operate under national standards specifying how clean the water that they discharge must be. The standards are contained in the National Pollution Discharge Elimination System (NPDES). The U.S. Environmental Protection Agency (EPA) and state environmental regulatory agencies are responsible for monitoring treatment plant operations and enforcing NPDES standards.

Federal funding for wastewater treatment works, once generous, has declined markedly in recent years. EPA grants for project construction have contributed more than $40 billion to the cause during the past fifteen years, but policy changes initiated in 1987 are currently phasing out this aid. At the same time, however, operating costs are growing rapidly. They amount to some $6 billion annually and constitute an even larger segment of the municipal wastewater budget than does the construction portion. Thus, spiraling demands triggered by regulatory standards face constrained state and local budgets, while federal financial support is being withdrawn. There is an urgent need for managerial innovations at the municipal level to cut the costs of wastewater treatment while maintaining or improving the quality of treatment processes.

In the face of the tightened regulatory standards and budgetary constraints, many municipalities are searching for methods of reducing operating costs while meeting their continuing, indeed escalating, clean water responsibilities. In such settings, private companies have increased their presence. Through a variety of arrangements—standard renewable contract operations, long-term commitments, and completely privatized deals—numerous cities have utilized private enterprise to assist with some of their managerial burdens.

Defenders of the practice of contracting out operations claim that it offers a means of tapping managerial and technical experience and expertise to improve performance while using market competition to lower operating costs. Critics are concerned that such arguments are exaggerated, that effluent quality is not likely to be a primary goal of the companies, and that contracting will prove to be more costly in the long run. Given the resource constraints and regulatory responsibilities associated with wastewater treatment, a systematic examination of the performance and operating costs associated with alternative arrangements (standard municipal management in conjunction with various types of contracting patterns) would obviously be useful.

However, despite the emerging importance of contracting for plant operations, no systematic investigations of this subject have yet been published. Some studies have appeared on such topics as the feasibility of comprehensive privatization (Gilbert and Miller, 1987; Heilman and Johnson, 1989). But thus far investigators have not systematically compared operating costs and performance of various contracting alternatives for treatment works.

Not surprisingly, given the importance of the topic, EPA recently utilized a consultant to compare public and private alternatives in this field. However, the analysis included only eight plants, did not distinguish among types of contract arrangements, made no effort to control for effluent limits or identity of contractor, surveyed only overall cost and performance figures rather than detailed subcategories, and focused only on the immediate period surrounding the shift from public to private management (U.S. EPA, 1987).

Since the value and advisability of contracting for government services are highly dependent on both context and service, it is not obvious whether the claims of contracting advocates have empirical merit in the field of wastewater treatment. The operation of WTWs differs in significant ways from many other kinds of governmental services: This policy sector is capital intensive, and knowledge of the relevant technologies is rather widely dispersed. The research reported here, by addressing this question systematically, is aimed at helping to deal with this important practical question.

Research Question

The research reported here utilizes information on the comparative operating costs and performance of municipal WTWs managed under different public and private arrangements. The scope of the project is a nationwide empirical examination of evidence from individual plants, municipalities, and regulatory agencies. The objective is to determine the effects on operating costs and on quality of operations of various kinds of contracting for plant management.

The question of the relative costs and performance of various options during operation is a difficult and important issue. Shifts in federal clean water policy (via the Water Quality Act of 1987) move significant responsibility to the states through the creation of state revolving loan funds (SRFs), the uses of which are mostly still to be determined by state officials. Yet, tightened federal regulatory standards place major constraints on local managers. Thus, the attractiveness of standard contracting for plant operation, as well as of more innovative and experimental options such as comprehensive privatization, may well hinge on the question of whether these arrangements make a difference as to cost and performance during operations. The answer to this question can have a substantial impact on the uses to which states put their SRFs, the managerial choices reached by municipalities, and the degree of compliance achieved by WTWs.

My discussions with industry representatives and local officials indicate that some believe plant operation by stable, reputable private management can improve service quality and reduce costs to the municipality. The advantage of private-sector flexibility, economies of scale, expertise, and experience might bring measurable advantages, especially to smaller and medium-sized communities. Nevertheless, as indicated earlier, these ideas remain controversial and largely untested. Other interested parties, especially AFSCME, view these arguments critically and suggest that outside contracting entails less obvious but ultimately significant operational costs (for example, AFSCME, 1983). Systematic efforts to address this topic are warranted.

In my research project, cost and performance data for a number of treatment works in various parts of the country are analyzed. Comparisons of two types are conducted: (1) across the several types of private-provision setting and (2) longitudinally, or over time, as service provision in some localities has shifted from public to private contract provision.

These comparisons, conducted on detailed data, address several, more specific questions. For example, advocates of contracting assert that long-term contractual commitments by municipalities allow companies sufficient predictability to avoid suboptimal, short-term decision making and provide personnel with sufficient stability to improve morale. Thus, the longer the contract, the lower the costs and the higher the overall performance. Adversaries of such deals assert that longer contracts provide incentives for "lowballing" (estimating costs on items such as maintenance at artificially low levels to obtain a bid, and subsequently negotiating for altered and more expensive contractual provisions). The present research project was designed to shed light on such disputed questions. The overall objective of the research is to connect the emerging institutional innovations in the field of wastewater treatment to the practical questions of cost and performance, so that policy analysts and government officials can make informed recommendations and decisions.

Research Design

The approach used in this project involves analysis of a large data set made available by a leading firm in the industry, face-to-face and telephone interviews to assess and strengthen the validity and reliability of the quantitative analysis; and limited use of an additional data base from EPA to make broader comparisons and thus deal with an aspect of external validity. Here, I describe the core data set, justify its use, and indicate some of the procedures followed in conducting the research.

One of the major firms involved in contracting for the management of treatment works operations, a relatively stable firm with extensive experience and a good reputation in the field, agreed to allow me access to its detailed data on both costs and performance regarding all of its WTW operations. The firm has developed operations experience in a range of contract settings and has also employed for its own management decision making a data base that includes considerable information on operations in each of the plants under company supervision. If notions about possible advantages of private contracting have empirical support, such support is likely to be found in the pattern of operations of such a company. Thus, the data under investigation here should provide a good test of the contracting question in this particular policy sector.

The data include cost figures in as many as 155 standard categories, characteristics of plant design and regulatory constraints, and thorough information on performance: effluent quality as measured by multiple indicators. (Because of apparent inconsistencies in methods for coding costs into detailed subcategories across different treatment facilities, the very large set of cost categories was reduced for analysis to a much smaller number that reflected more reliability in the data.) Furthermore, for those facilities that had previously operated directly under municipal management, cost and performance data from earlier years were recorded in a fashion roughly comparable to the firm's information about its own operations. I supplemented these data on the individual plants, their characteristics, cost, and performance, with additional information on the contracts negotiated with individual locales, the degree of oversight exercised by cities over the contractor, the local political and administrative situation, and other potentially important features of the individual cases (for example, whether the labor force is unionized, whether unusual local conditions obtain).

Data of some sort were available for approximately eighteen contract settings representing all regions of the United States. (This number was reduced to fifteen for much of the detailed analysis because of data quality problems and/or insufficient operating experience in a few locations.) These facilities represent several types of treatment technology, community size, plant size, and contract length. Two cases of comprehensive privatization were represented in the data analyzed, as were several unionized plants.

The use of this kind of information for a systematic inquiry into the research question addressed by this study presents significant advantages. For instance, standard reporting procedures and a single information system within the company increase the likelihood that data from different WTW facilities can be compared. Furthermore, examination of the entire set of plants operated by a leading firm provides a fair test of the cost and performance issues, while also holding constant some of the most important variables that can affect WTW operations: All plants in the sample are run under a single management philosophy, and the level of corporate expertise being brought to bear on different WTW operations is relatively constant. Of course, it is also true that one firm may not be fully representative of the contract operations industry overall; thus, the issue of external validity is raised by the basic design used here. This matter was addressed through certain refinements of the research design, as explained below.

Understandably, the company that agreed to make these data available for independent analysis had concerns about the release of proprietary information such as detailed cost data. Precautions were taken to accommodate both the company's appropriate concerns and the basic need for independent research. During the course of the research, the firm provided all available requested information. Full access was obtained to the data bases of the company. Top management officials were interviewed at length regarding contract operations; visits, inspections, and interviews were permitted at several company plants of my choosing; and various relevant files and records of company business were provided upon request.

Since the bulk of the quantitative analysis at the heart of this research was conducted from privately gathered data in the possession of a single company, questions could readily be raised concerning the reliability and validity of the data, and the credibility of the interpretation. Therefore, several steps were undertaken to strengthen the significance and practical value of the findings:

1. Company headquarters officials were interviewed in depth in regard to how they generated and used their data, and in regard to the conduct of contract operations.

2. Detailed field visits were made to four WTWs managed by the company, and briefer visits were made to three other facilities in the data set. I chose all of these sites on the basis of their importance for the research questions under investigation. In three of the four principal sites, the communities had managed their own WTW operations for a period of time and had then contracted directly with the firm. In the fourth case, the city had managed its operations for a time, had then contracted with another firm, and had more recently switched contracts to the company whose data were analyzed in this study. Three of the major site visit cases were also included in the EPA consultants' study mentioned earlier. The cases selected for visits were also chosen to represent different regulatory environments; for instance, the four sites are located in four different states and EPA regions. The company permitted inspection of facilities, interviews of field personnel, and examination of local data-recording routines. These steps were designed to assess validity and reliability as well.

3. At the time of the field visits separate interviews were conducted

with municipal government managers in the same locales. The primary objects here were threefold. First, this step allowed an assessment of the reliability of data on company costs and performance. Second, information on municipal costs and performance during the period prior to contracting was obtained. And, third, city officials were given a chance to provide their candid assessment of the advantages and disadvantages of contracting in this policy sector, based on their own experiences.

4. Interviews were conducted with representatives of a small sample of additional companies in the WTW contracting industry to assess the generalizability of the results of the present research. One interview was held at a corporate headquarters; the others were conducted by telephone.

5. Important plant and performance characteristics in this sample were compared with similar characteristics of a sample of facilities recorded in PCS, one of the national data bases of EPA. The data base covers facility-level performance, design, and compliance data for virtually all municipal treatment works. While operating cost data cannot be assessed through this data set, performance measures definitely can be. This step was taken to address one important aspect of the question of external validity: determination of how generally the results of the company-based study apply to other municipal WTWs.

Research Results

The unit of analysis for this study is the municipality and its WTW. As indicated earlier, the analysis of both costs and performance can be conducted both across plants or locales (in other words, cross-sectionally), as well as longitudinally. The presentation of results here is ordered in that sequence.

Comparing Contract Operations in Different Settings. The research in this study includes comparisons across various contract operations settings, both to determine the range of variation across locales' experiences and to seek explanations for differences in both costs and performance of operations in different privately operated cases. Within the experience of this one company, all of whose contract operations continue to the present, there is considerable variation in certain characteristics of local settings; in other respects, however, there is consistency across the sample.

The local jurisdictions that decided to contract their operations are located throughout the country, with a substantial number clustered in the region that contains the headquarters of the company. The locales range from very small communities (populations of a few thousand) to medium-sized cities (populations of approximately one-hundred thousand). Indeed, small and medium-sized cities have thus far constituted the general market for contract operations. All plants in the sample are major facilities, that is, they are built to process at least one million gallons per day (MGD). In this sample, facility design sizes range from one to thirty-one MGD. All but two of the plants are in the range one to ten MGD, a typical size for major facilities. Analyses were performed both on the entire set of plants and also on those of size one to ten MGD to test for patterns in operations within that more restricted group. Some of the cases involve old plants (twenty years or more), while others are very new. Plant technology ranges from nearly outmoded to highly innovative.

The local wastewater problems and politico-administrative arrangements also vary considerably within the sample. The influent (sewage to be treated) ranges from light residential to heavy industrial; some plants employ unionized labor forces, while others do not; the communities are located in states that differ greatly in the stringency and enforcement of their regulations; in some communities, clean water and wastewater matters are very salient political issues, while in others the issue is invisible; and some local governments monitor and oversee their contract operations closely and vigorously, while in other settings the local government rarely considers the issue—indeed, some locales adopted the practice of contracting precisely to avoid the set of concerns associated with WTW operations.

The contracts negotiated by the firm with the individual communities contain a number of variations as well. Obviously, the price charged for the operations service is determined in individual settings. Indeed, the company routinely begins detailed negotiations on this subject only after it sends a specially selected "pricing team" into a community for an intensive examination of the city's facilities, technology, and wastewater conditions. In some locales the company has been in charge of operations for as many as eight years already, while in others the public decision to contract has been much more recent. Contracts also vary considerably by length (between one and twenty-five years), with the modal period being five years. Another individually negotiated element is the contract clause(s) on "savings." In most of the cities examined, contracts in force contain a provision that budgeted dollars not spent at the end of the fiscal year are to be split between locality and company on the basis of some formula; the terms of the formula vary somewhat from jurisdiction to jurisdiction. In the comprehensively privatized cases, no such clause is a part of the contractual arrangement, since the facility is totally owned by the company, which controls both costs and savings.

On other dimensions, however, the cases are substantially similar. For instance, several features of the contractual agreements with the local governments are basically constant. The company routinely agrees to hire the entire municipal labor force upon assuming managerial control over the locale's wastewater treatment facilities, accepts full legal responsibility for compliance with regulatory limits, and applies a companywide equal employment opportunity policy in personnel decision making. The contracts also typically include a price escalator tied to the cost-of-living index or else allow for annual price adjustments in a setting where both parties agree that "operating costs depend greatly on the amount of flow and load to be treated" (to borrow language from one such contract). In short, for a number of potentially interesting contract issues, there is little variation in the sample and thus no opportunity to test for impact on actual costs and performance. In sum, the independent variables explored in this research include the characteristics of locale, plant, and contract that might plausibly have an impact on costs and performance.

Costs. One major set of dependent variables consists of various costs for contract operations. They were recorded from the company's data bases. Analysis is largely restricted to overall costs and the major cost categories: total labor costs (rather than, say, employee benefits, overtime, and so on), total utility costs, total chemical costs, and total repair and maintenance costs. In addition, because most of the contracts were written to exclude any incentive for the company to save budgeted maintenance dollars, another variable, representing overall costs less maintenance expenses, was created. (One-hundred percent of the savings in this cost item typically revert to the locality; the clause encourages the firm to invest in preventive maintenance, thus possibly saving on other cost items over the long run.)

Furthermore, since it was expected that a major factor determining treatment costs in these various categories is the scale of operations, another set of cost variables was created to standardize the cost items by plant size. Thus, labor costs in a given year per MGD of plant design size can be examined. Costs were recorded for all contract operations over the period 1981–1988. However, since the data are most reliable and complete for recent years, and since the company had more locales under contract in those recent years, analysis is concentrated on fiscal years 1987 and 1988.

What can be said about the determinants of contract operations costs, at least for this one company's set of plants in the recent period? First, the cost measures used in this study are highly intercorrelated with each other. Labor costs and repair costs, for instance, tend to be higher in facilities where chemical and utilities costs are also high. The intercorrelations hold not only when costs are measured directly in dollars but also when costs are standardized by plant size; higher-cost facilities per MGD in one category are likely to be higher cost in other categories. These relationships are present in both years 1987 and 1988, with Pearson's r's of between .60 and .90; virtually all correlations among standardized costs are statistically significant at the $p < .05$ level.

A second notable feature is how many plant, locale, and contract features are *unrelated* in a statistical sense to the various measures of plant costs. Although the costs themselves vary widely, the difference in general is not explained by items such as the following: geographical location, stringency of state policy on clean water matters, degree of EPA-delegated authority to the state regulators, history of the facility (that is, whether it began as a publicly operated plant or was initiated as a contract operation), visibility of clean water issues in the locale, influent quality or effluent limits, contract length for privatization, formulation of the "savings" clause in the contract, age of the plant, or type of plant technology.

These "negative" findings are very interesting and are consistent with the company management's view on costs and pricing: Many idiosyncratic features of the local situation (for example, quality of the municipal work force, extent to which the physical facilities have been maintained, condition of pump stations and other structures unique to particular systems, engineering details of a particular plant, and topography) drive costs. And absent an intensive study of all features of a local situation, there is seldom a simple explanation or prediction for the various types of costs that arise in contract operations. In short, cities looking to the experience of other locales for a clear sense of what a "fair" cost may be for contract operations are unlikely to find clear answers from cross-sectional analysis.

Furthermore, these findings are largely consistent with the firm's claim that some details of a contract's clauses (for example, contract length, type of savings clause) have a negligible effect on costs to a city of contract operations. The company argues that the structure of the contract does not really provide incentives to plant managers to shift costs in particular directions, and that spending decisions are determined by engineering and plant (that is, technical) needs.

Additional support for this line of argument derives from an analysis of the company's use of performance awards for its plant managers. For this particular firm, one element of the incentive structure operating on company personnel is the possibility for plant managers to obtain annual bonuses. Plant managers have considerable discretion to encumber costs for the company. Performance awards are keyed to savings from budgeted amounts for a particular plant. However, analysis shows that the size of performance awards provided to managers is not correlated with plant costs, either dollar costs or standardized costs. This lack of relationship holds for both 1987 and 1988. Furthermore, to see whether cost-*cutting* or -*containment* from one year to the next is rewarded in the performance bonuses, the differences between 1988 and 1987 costs were analyzed with respect to performance awards for 1988. Again, no significant relationships were found to exist. The size of managers' performance awards is also unrelated to the scale of their respective jurisdictions, whether measured in total costs, plant size, or staff size. Thus, the frequent claim among students of administration that bureaucracies distribute rewards disproportionately to those who build or preside over large "empires," rather than to those who perform well, seems unsupported in the case of this contract operations company.

Nevertheless, there are interesting patterns in the data. Some variables are related to plant-operating costs. As one would expect, plant size is highly correlated with almost all of the unstandardized costs (Pearson's r is .89 for 1987 costs, .88 for 1988; both significant at the $p < .05$ level). Indeed, because operating costs are so closely tied to plant size, all cost variables were standardized by plant size for further analysis.

Examination of these standardized costs provides a better sense of what factors might contribute to relative efficiency. Bivariate analyses (simple correlations and analyses of variance) reveal two interesting patterns of relationships. Of the wide array of locale, plant, and contract characteristics examined, plant design size and the original starting date of the service contract are consistently and significantly related to almost all of the standardized costs for each of the two years that were investigated.

An examination of the bivariate relationships between plant size and the various standardized cost measures suggests a nonlinear pattern: a negative slope decreasing in absolute value and approaching zero at high levels of the independent variable. The pattern is consistent across various measures and also makes substantive sense. Such a relationship with plant size suggests economies of scale that gradually diminish in larger plants. An exponential function was tested for its ability to "fit" or explain the variance in standardized costs. The function produced a very close and statistically significant fit to the data.

In addition to the bivariate relationships tested in this study, stepwise multiple regression was utilized. Independent variables selected for testing were chosen on the basis of both the bivariate analyses and the theoretical expectations derived from literature on contract operations. However, the multivariate results did not change significantly the conclusions reached through the bivariate analysis. The main conclusions, some fairly tentative because of the limited sample size, are as follows.

One of the clearest findings is that larger plants are associated with lower costs per MGD. This relationship holds when the analyses are performed with or without the two largest (that is, greater than ten MGD) facilities. Apparently, the economies of scale can be seen among major facilities even at sizes at or below ten MGD. This finding has practical consequences. For instance, coordination between or among local communities in a joint contract arrangement to operate a larger facility may lower costs, at least during operations. It may be worthwhile for communities considering new construction of a WTW, therefore, to explore cooperative arrangements with neighbors and to solicit cost estimates for facilities of various sizes.

Whereas the relationship between plant size and standardized costs is easily explained with the familiar concept of scale economies, the tentative finding with respect to contractual starting date is more intriguing. The results indicate that in this sample the earlier a locale initially contracted with the firm for services, the higher the operating costs *in recent years,* when compared with other plants under contract in recent years. Several potential explanations for this relationship were examined, but none was supported by the data. For example, earlier contracts might be associated with older plants; thus, higher costs could be explained by deteriorating facilities. However, length of contract period is not related to plant age; and age is unrelated to any cost measures. Nor is the finding explained by such variables as plant technology or effluent limits, either of which might be expected to be linked to the date of original contracting.

In short, the finding seems linked to the length of the contract period alone. The data available, and the design of this project, do not allow an unambiguous interpretation of this finding. However, several explanations are possible; I review here two that illustrate the subtleties and complexities of service contracting.

In recent years we have seen heightened competition on grounds of cost in this industry. Thus, locales initiating contracts in more recent years may have been able to take advantage of the more well developed market effects on the supplier side. Certainly, this explanation fits both with the oft-claimed benefits of contract services (that is, competition lowers costs) and with the view of the industry offered by representatives of this firm and others with which it competes. There are now more companies, more options for cities to consider, and consequently more cost advantages to newer contracting sites than previously existed. Still, it is not altogether clear why "older" contracting sites do not seem to reap the advantages of lowered costs when they renew their contracts in the more competitive current environment. Certainly, at minimum city officials may want to become educated about the direction of industry pricing patterns as they approach a recontracting decision.

A second possibility has to do with an issue sometimes raised by labor representatives in debates over private service provision. The claim is that contract terms are initially drawn in a fashion that seems advantageous from the point of view of a local government. In a few years, however, when the city has eliminated its own labor force and expertise in the service area because of the contracting option, the company is in a more advantageous bargaining position. Recontracting then may occur at terms less favorable to the locale. Although this kind of influence may be at work in the sample examined for this study, there is no direct evidence. Analysis of variance shows no relationship between standardized costs and the recontracting experience of a locale: Cities in their second or third contracting cycle experience no higher costs than do cities currently operating under the initial contract with the firm. Indeed, local officials interviewed at the sites visited for this research uniformly expressed satisfaction with their contract arrangements.

In any event, the pattern of relationships between costs and length of a community's association with the contract firm contains lessons for public managers. Those considering the option of contracting should be alert to the trends; they might take these findings into account when they consider the structure of escalator or negotiating clauses and when they approach a recontracting date.

One additional result of interest in the regression analysis is that the degree of oversight by the locality over its contractor is statistically related to several standardized costs. This finding is substantively provocative: More vigorous oversight may be associated with lower standardized costs. The interpretation here is straightforward and potentially quite valuable.

Performance. The other important class of dependent variables used in the cross-sectional analysis is a set of measures of performance. Different localities face different wastewater problems; thus, it is difficult to develop an unambiguous measure of performance. However, the best indicator is probably a plant's ability to meet regulatory standards.

Communities typically are required to adhere to a series of such standards, each of which refers to a different type of pollutant. Those analyzed for this study are the two most frequently cited measures: the five-day biological oxygen demand (BOD5) and total suspended solids limits. In virtually every case the firm's plants met permit limits in all or almost all of the months during the two-year period under review. In several instances, the plants were substantially below permit limits on a regular basis. The data are clear: In recent years the firm has consistently performed well in all sites, regardless of local conditions, contract features (within the range of variation contained in the sample), and plant characteristics. Contract operations, at least with this firm, have produced regulatory compliance. This finding is also a source of great satisfaction and relief in many of the communities under contract. In the four visited for this study, there was enthusiastic support for the performance of the firm in recent years.

Some of the contract communities in the sample had not always experienced such high levels of performance under earlier conditions. This observation raises the important question of whether or not performance success is a likely result of private operations *per se*. In other words, does this firm (and, possibly, do other companies) outperform governments that manage their own treatment works? To address this question, as well as the matter of public versus private costs, one needs to analyze data not across contract facilities but rather between public and private settings. The next section presents the results of such comparisons conducted in two fashions: by examining some current contract plants longitudinally to compare public and private periods and by comparing the firm's plants with a larger sample of publicly operated facilities.

Comparing Public- and Private-Sector Provision. This investigation included a comparison between public and private periods of service provision for the WTW facilities in the sample. For a variety of reasons only four cases were suitable for this analysis. Three of the four public-to-private cases were visited. One city that had employed another firm before changing to its current contractor was also visited; the information gained during the interviews there provided some data on the range of contracting experiences possible for a community. Though limited, these data do cast some light on the issues that arise in contracting and in its analysis and evaluation.

Costs. With regard to costs over time, an unambiguous trend cannot be documented from the data. In two of the locales overall costs declined—in one by a slight amount, in the other by a larger amount—at least in the period immediately following contract initiation. (Of course, as the earlier discussion indicates, a longer relationship with the firm is associated with rising costs.) In one city visited, costs after contracting were slightly higher than during municipal management. This set of mixed findings is consistent with the results of the EPA-contracted study cited earlier (U.S. EPA, 1987). However, these simple comparisons may obscure as much as they reveal. A locale may be underspending on its own wastewater treatment operations, thus gradually generating a compliance or a labor problem, and then find itself in need of outside assistance. In such a case, cost increases at the time of contracting may say nothing about relative public-private efficiency. Contract firms can point to instances of savings, and opponents of contracting can identify at least some cases where contracting may have cost more than public operations even in the short run. The earlier analysis, as well as the information gathered from detailed interviews in the various cities for this study, suggest that the likelihood of savings from contracting varies greatly from case to case.

It should be noted that contract operators, including the firm examined closely in this study, point to certain sources of efficiency that may plausibly be present. These fall largely into two categories: market-induced elements and scale economies. The former may derive from private-sector operation per se; the latter may also be available to larger communities whether the facilities are publicly or privately managed.

With regard to the former, firms in this business point to the incentives for efficient operation present in market settings and absent from government. Market forces, it is claimed, encourage companies to apply the correct amount of chemicals, operate efficiently with regard to energy, greatly enhance training and improve salaries while slowly generating labor savings through attrition, and produce increased savings and profits in the long term by a vigorous program of preventive maintenance. Industry representatives believe that many local governments, insulated from the stimulus of the market, tend to react slowly and suboptimally as they manage their own operations. Several plant managers with both public- and private-sector experience who were interviewed for this study commented along these lines as they compared the different varieties of operations. However, it is simply not possible with the data at hand to assess conclusively the strength of market forces as they affect plant costs over the long run.

The company examined in this study also unquestionably can bring certain scale economies into play, even in small plants: In newly contracted facilities the firm can apply management systems developed and perfected elsewhere; can use its own copyrighted computer software for such purposes as process engineering, maintenance scheduling, and cost accounting; and in any facility can draw on its national staff of experts in various phases of WTW operations to solve nearly any problem, often without additional cost to the locale. Evidence for the value of this latter set of impacts was abundant during field and headquarters visits conducted for this study.

Performance. Costs, however, tell only one part of the story. Also significant is the question of performance. None of the cities examined during the course of this project had contracted primarily to save money. In all cases, some other force was more important: labor or managerial difficulties, technological problems, and/or compliance. It is thus crucial to determine the relative performance of these plants in the public and private periods if one is to assess fairly how successful the contracting has been for the communities involved.

The answer to the performance question is clear for this sample. For the cities with prior public operations experience, contracting is associated with substantially improved performance. Currently, virtually all of the company's plants consistently meet permit limits. Indeed, six of the firm's facilities have received state, federal, and professional awards for quality of services. Although some of the sites have had occasional brief difficulties with a permit limitation, none can be considered a "problem" facility; none, for instance, is currently under sanctions, moratoria, or other less formal regulatory actions by EPA or the various state agencies. In fact, the reputations of many of these plants and their management are quite high within the states.

This pattern is much different from that found in these same cities prior to contracting. At the time that the option of contracting was first considered in these locales, most of the communities were experiencing regulatory pressure, some were under sanctions and moratoria, and many were having difficulty establishing a plan for resolving the immediate problems. In one classic instance within the sample, a local plant that had been out of compliance for three years finally contracted with the firm; within twenty-one days the plant met its permit standards and has not been out of compliance in the several years since that time. The company sent more than thirty wastewater experts to the field site for a few weeks, assessed the situation, redesigned portions of the facility, and solved the performance problem.

Is good performance an inevitable result of private operations, and is questionable performance the likely result under public management? It is clear that the answer to both of these questions is no.

Private operations can, but do not always, bring quality. Although this firm currently performs well for its clients, not all prior experiences were so unambiguously positive. In one site visited during this study, for instance, the locale is now quite pleased with operations but had serious concerns several years ago. During the early years of its contract with the community, the company had not brought the facility into compliance with federal regulatory standards, and the quality of plant management was suspect. The locale threatened suit. The company then worked with the community to solve the problems, and all parties now express satisfaction. The local officials learned from this experience that contract operations are likely to work well only when the community does not abdicate its role regarding oversight and accountability. The locality maintains experts on engineering, management, and finance on its oversight board; it takes an active stance with respect to such matters as operations at the facility, details of the plant budget, a variety of reports required regularly, and performance specifications developed locally. Both parties to this arrangement describe the setting as an instance of true public-private partnership. This plant was the recipient recently of the regional EPA Best Operated Plant award in the facility's size category.

Some of the other locales visited for this study do not participate in as vigorous a program of oversight as pursued by the particular community described above, and they too have avoided significant recent performance problems. However, one of these locales had less pleasant experiences with its previous private operator; the result was a change in contractors after a few years. The argument for active oversight gains strength from the findings of the cross-sectional analysis presented earlier: The data suggest that more oversight may be associated with lower costs. Thus, from the perspective of communities considering the contracting option, active though not intrusive oversight may provide advantages on both the cost and performance sides of the equation.

Comparison of the plants in this study to a larger sample of publicly operated facilities provides some information about the question of whether publicly managed WTWs in general are prone to the kinds of problems experienced in this smaller set. In a sample of major facilities in the southeastern United States (EPA Region 4) built between 1984 and 1986 ($N = 24$), only two facilities consistently exceeded permit limits during a recent year (1987). Most remained consistently in compliance. Three conclusions can be drawn. First, it is quite possible in many communities for public operations to work satisfactorily. Second, the sample of cases in this company's data base are unusual: Thus far, most cities that have contracted have done so at a time when they are facing rather severe difficulties; and the majority of cities whose facilities are in compliance have not thus far seriously considered contracting. (This evaluation is shared by representatives of other companies in the industry as well.) Third, contracting does not automatically improve performance for all cities; important factors to consider in the decision are the structure of public-private relations as articulated in the contract, the experience of the firms under active consideration, and the state of local public management.

Conclusion

This study contains an assessment of one mode of alternative provision for one public service: contracting for operations of municipal WTWs. The evidence indicates that performance advantages may be possible for certain types of communities when contracting is sensibly employed.

Conclusions about the financial implications of contracting are necessarily more difficult and tentative. Contracting is unlikely to be a fiscal panacea. Indeed, to the extent that a community considers only the cost side of the ledger, it may pay for its shortsightedness with compliance problems, labor difficulties, or political discord over the longer term.

There is an implicitly optimistic theme to the analysis of contract operations costs presented above, however. At least within the sphere of this company's experience, some communities obtain lower standardized costs at no sacrifice to output. Furthermore, the variables associated with these savings seem to be items over which local government managers can under certain circumstances exercise control. Savings are possible in larger facilities and when the local government assumes an active and energetic oversight role. The pattern of cost relationships with plant design size, with length of contractual association, and with degree of active oversight points to elements of a contractual relationship that are under at least some direct influence by the parties involved.

The most general lesson to be learned from this study, then, is familiar. Effective and efficient provision of public services necessitates active and informed public management—even, or perhaps especially, when *private* actors are also directly involved.

References

American Federation of State, County, and Municipal Employees (AFSCME). *Passing the Bucks: The Contracting Out of Public Services.* Washington, D.C.: AFSCME, 1983.

Armington, R. Q., and Ellis, W. *This Way Up: The Local Official's Handbook for Privatization and Contracting Out.* Chicago: Regnery Gateway, 1984.

Borcherding, T. E., Pommerehne, W. W., and Schneider, F. "Comparing the Efficiency of Public and Private Production: The Evidence from Five Countries." *Journal of Economics,* 1982, Supplement 2, 127–156.

Brooks, H., Liebman, L., and Schelling, C. S. (eds.) *Public-Private Partnerships: New Opportunities for Meeting Social Needs.* Cambridge, Mass.: Ballinger, 1984.

Brudney, J. L. "Public Versus Private in the Delivery of Services." *Urban Affairs Quarterly,* 1984, *19* (4), 550-559.

DeHoog, R. H. *Contracting Out for Human Services: Economic, Political, and Organizational Perspectives.* Albany: State University of New York Press, 1984.

Fitzgerald, M. R., and Lyons, W. "The Promise and Performance of Privatization: The Knoxville Experience." *Policy Studies Review,* 1986, *5* (3), 598-605.

Florestano, P. S., and Gordon, S. B. "Public vs. Private: Small Government Contracting with the Private Sector." *Public Administration Review,* 1980, *40* (1), 29-34.

Gilbert, A., and Miller, J. "Privatization of Wastewater Treatment Facilities: Making the Decision." Unpublished manuscript, J. F. Kennedy School of Government, Harvard University, 1987.

Heilman, J. G., and Johnson, G. W. "A Feasibility Study of the Privatization of Public Wastewater Treatment Works." Report prepared for the U.S. Geological Survey, Department of the Interior, Award No. 14-08-0001-G1288, January 1989.

Kolderie, T. "The Two Different Concepts of Privatization." *Public Administration Review,* 1986, *46* (4), 285-291.

McDavid, J. C. "The Canadian Experience with Privatizing Residential Solid Waste Collection Services." *Public Administration Review,* 1985, *45* (5), 602-608.

Marlowe, J. "Private Versus Public Provision of Refuse Removal Service: Measures of Citizen Satisfaction." *Urban Affairs Quarterly,* 1985, *20* (3), 355-363.

Palumbo, D. J. "Privatization and Corrections Policy." *Policy Studies Review,* 1986, *5* (3), 598-605.

Perry, J. L., and Babitsky, T. T. "Comparative Performance in Urban Bus Transit: Assessing Privatization Strategies." *Public Administration Review,* 1986, *46* (1), 57-66.

Savas, E. S. "How Much Do Government Services Really Cost?" *Urban Affairs Quarterly,* 1979, *16,* 23-42.

Savas, E. S. *Privatization: The Key to Better Government.* Chatham, N.J.: Chatham House, 1987.

Slawsky, N. J., and DeMarco, J. J. "Is the Price Right? State and Local Government Architects and Engineer Selection." *Public Administration Review,* 1980, *40* (3), 269-274.

Starr, P. *The Meaning of Privatization.* Project on the Federal Social Role, Working Paper No. 6. Washington, D.C.: National Conference on Social Welfare, 1985.

Stevens, B. J. (ed.) *Delivering Municipal Services Efficiently: A Comparison of Municipal and Private Service Delivery.* Technical report. Washington, D.C.: U.S. Department of Housing and Urban Development, Office of Policy Development, 1984.

Straussman, J., and Farie, J. "Contracting for Social Services at the Local Level." *Urban Interest,* 1981, *3,* 43-50.

U.S. Advisory Commission on Intergovernmental Relations. *Intergovernmental Service Arrangements for Delivering Local Public Services: Update 1983.* Technical Report A-103. Washington, D.C.: Department of Housing and Urban Development, 1985.

U.S. Environmental Protection Agency (EPA). *Contract Operation and Maintenance.* Washington, D.C.: EPA, Office of Municipal Pollution Control, Planning and Analysis Division, 1987.

Valente, C. F., and Manchester, L. D. *Rethinking Local Services: Examining Alternative Delivery Approaches.* Technical Report No. 12. Washington, D.C.: International City Management Association, Management Information Service, 1984.

Weimer, D. L., and Vining, A. R. *Policy Analysis: Concepts and Practice.* Englewood Cliffs, N.J.: Prentice-Hall, 1989.

Wollan, L. A., Jr. "Prisons—The Privatization Phenomenon." *Public Administration Review,* 1986, *46* (6), 680-681.

Laurence J. O'Toole, Jr., is professor of political science at Auburn University, Alabama. He is author, co-author, or editor of four books and more than forty articles and book chapters.

*States can manage the solid waste stream in part by using
financial incentives that encourage the beverage industry to
undertake voluntary recycling programs. Two approaches, in
particular, represent innovations in public-private partnerships, or
privatization: preemptive partnerships and compelled partnerships.*

Using Financial Incentives to Manage
the Solid Waste Stream

Charles J. Spindler

Solid waste disposal has the attention of government, industry, and the
public. The litany of the solid waste dilemma is all too familiar: Solid waste
generation continues to escalate, packaging proliferates; landfills continue
to close, while sites for new landfills and incinerators are scarce resources.
Incinerators are suspected of fouling the air with hazardous emissions, and
landfills are known to have fouled water supplies. New federal requirements
to control landfill hazards will significantly increase the cost of landfill
disposal. The interstate transfer of waste is a vigorously contested issue for
some states. Local governments bear most of the increasing financial burden
of the solid waste disposal; in turn, the costs are passed along to the public.

State and local policies for solid waste recycle/reuse and waste reduction are evolving rapidly. Policies directed at reducing the postconsumer
waste stream extend across a continuum that includes voluntary initiatives
such as educational programs and voluntary source separation, technical
assistance to industry, and financial incentives for industry recycling such
as tax breaks, credits, and loan and grant programs. The policy continuum
extends in some states to compulsory initiatives such as mandatory source
separation and recycling programs, litter abatement laws, requirements for
reusable containers, mandatory deposit fees, material taxes, prohibition of
selected materials, and the exclusion of waste from disposal based on
origin or type of waste.

This chapter develops a model to categorize public policies directed
at diverting postconsumer waste from the waste stream, and industry initiatives in the context of these policies, as a step in evaluating the effectiveness of those policies. An enhanced understanding of the impact,

effectiveness, and limitations of these policy approaches should contribute to an increased ability to control the solid waste stream.

First, I discuss broad categories of public solid waste policies and industry responses. I then explore the development of a new category of public solid waste policy and industry responses. In this emergent form of public-private partnership, "preemptive" and "compelled" partnerships are examined. They represent a kind of "shotgun marriage" between the public and private sectors in solid waste management. How this union takes place, and its potential for long-term success, are examined. It is suggested that the model developed has applications in the study of other public policy domains.

The state of Florida's use of financial incentives to manage postconsumer waste (in particular, packaging waste) serves as an example of state-compelled partnerships. In 1988, Florida adopted a comprehensive approach to solid waste management that includes the delayed imposition of mandatory deposit fees on containers. Despite industry opposition to deposit legislation in general, the legislation in Florida received little opposition when introduced; the Florida Soft Drink Association issued a position paper expressing full support (Galvin, 1989). An explanation of why Florida's current legislation met with little resistance from the beverage industry is offered here. I also discuss recommendations made to the Florida legislature on the effects of packaging on the Florida waste stream (Spindler, 1989). Finally, I suggest directions for additional research.

Growth of Waste Sets the Public Policy Agenda

The introduction of the Uneeda Biscuit package in 1899 signaled the beginning of the packaging revolution (U.S. Environmental Protection Agency [U.S. EPA], 1971). Packaging has since evolved beyond its original purpose of protecting the product from the rigors of the environment. That is, packaging has become a potent marketing tool: it advertises contents, presents manufacturers' logos, and diversifies product lines by offering different sizes and container types.

Because of widely recognized problems of packaging disposal, the packaging revolution has not been universally acclaimed. Packaging constitutes a significant proportion of the municipal waste stream (MWS). In 1986, containers and packaging were estimated to comprise 30.3 percent (42.7 million tons) of the waste stream and were projected to be 30 percent (50.7 million tons) in the year 2000 (U.S. EPA, 1988). These figures represent net discards after material recovery. Approximately 10 million tons of containers and packaging were recovered from the MWS in 1986; corrugated containers constituted 8 million tons of the total packaging recovered (U.S. EPA, 1986).

The disposal problem created by packaging is simple. Packaging that protects the product from the environment continues to resist the environment long after the product is consumed. Packaging buried in landfills remains relatively intact for decades; new landfill requirements that effectively seal landfills will almost completely halt packaging deterioration.

No other sector of industry has received as much attention or fostered as much debate over the environmental impact of packaging disposal as the beverage industry. Much of the attention directed at the beverage industry is the result of the industry's switch from returnable to nonreturnable packaging.

All beer and soft drink bottles were refillable and covered by bottlers' deposits until the 1930s and 1947, respectively (U.S. General Accounting Office, 1980). Manufacturers routinely reused their bottles to keep production costs low; the cost of bottle production was reduced each time a bottle was reused. In fact, there were no government-mandated deposits on bottles prior to 1953 and between 1957 and 1971.

As indicated in Table 2.1, in 1963 the ratio of returnable to nonreturnable glass beer bottles was 2.88 to 1; it was nearly 27 to 1 for glass soft drink bottles. But, as indicated in Table 2.2, by 1987 these ratios were reversed: The ratio of nonreturnable to returnable glass beer bottles was 36 to 1, and nearly 81 to 1 for glass soft drink bottles (Galvin, 1989).

Oregon often receives credit for enacting the first mandatory-deposit legislation in the United States in 1971. However, the earliest legislation of this type was adopted in Vermont in 1953, prohibiting the sale of beer or ale in nonreturnable glass containers (Jeffords, 1978). The intent of this

Table 2.1. Return Rates for Glass Beer and Soft Drink Bottles in Terms of Percentage Market Shares: Select Years, 1963–1975

	Returnable		Nonreturnable		Ratio	
Year	Beer	Soft Drink	Beer	Soft Drink	Beer	Soft Drink
1963	46	89	16	3.3	2.88	26.97
1967	35	65	21	13.0	1.67	5.00
1973	19	35	21	29.0	.90	1.21
1975 (est)	12	25	13	21.0	.92	1.19

Note: Ratios are returnable :: nonreturnable glass bottles.

Source: Based on Organization for Economic Co-Operation and Development, 1978, pp. 19–20.

Table 2.2. Return Rates for Glass Beer and Soft Drink Bottles in Terms of Billions of Units: 1983 Versus 1987

	Returnable		Nonreturnable		Ratio	
Year	Beer	Soft Drink	Beer	Soft Drink	Beer	Soft Drink
1983	.96	.18	13.63	8.35	14.20	46.39
1987	.33	.11	11.90	8.89	36.06	80.82

Note: Ratios are nonreturnable :: returnable glass bottles.

Source: Based on Galvin, 1988, p. 43.

legislation, like most of the deposit legislation to follow, was to reduce litter. The imposition of mandatory deposits in Vermont was vigorously opposed by industry. "The beverage and container industries, setting a pattern which would be repeated in the future, launched a massive campaign against it" (Jeffords, 1978, p. 13). The act was allowed to expire in 1957, in the face of massive lobbying. This early skirmish marked the beginning of the larger war between bottlers and government. Today, the beer, soft drink, and bottling industries remain almost totally opposed to mandatory deposits.

Overall, twenty-six states have adopted recycling legislation, but only nine states have mandatory-deposit fees in effect (National Solid Waste Management Association, 1989, p. 2). Since the Oregon act took effect in 1972, eight other states have enacted mandatory-deposit legislation for beverage containers (effective date shown in parentheses): Connecticut (1980), Delaware (1982), Iowa (1979), Maine (expanded in 1978), Massachusetts (1983), Michigan (1978), New York (1983), and Vermont (1977). Passage of mandatory legislation of any nature is difficult to achieve in the face of industry opposition. National mandatory bottle-deposit legislation has been introduced every year in Congress since 1977, without success (U.S. Congress-Senate, 1978). Since 1972, some 2,000 deposit proposals have failed (National Soft Drink Association, 1978, p. 3).

The adoption of modified forms of deposit legislation in California in 1986, and in Florida in 1988, are the only successful statewide efforts since the adoption of deposit legislation by New York in 1982. California requires the beverage industry to pay the state a redemption value on each beverage container sold in the state. Private recycling centers collect containers from consumers and fees from the state. The Florida system encourages voluntary recycling by industry under a threat of imposition of mandatory deposits if state recycling goals are not met. The Florida legislation marks a new approach to planned waste management in the United States: voluntary industry recycling to avoid the possibility of explicit governmental intervention in the market. This is the basis of a *compelled partnership*. This approach is not new, especially in Europe. The governments of Denmark, Germany, Norway, and Ontario, Canada, have attempted to achieve "voluntary" cooperation from industry on recycling under the threat of governmental solutions (see Organization for Economic Co-Operation and Development, 1978, pp. 145–147).

Techniques to Divert Postconsumer Waste from the Waste Stream

Efforts to divert postconsumer waste from the MWS can be categorized on the basis of initiating agent (government or industry) and action (voluntary or compulsory), as shown in cells 1 through 4 in Table 2.3. Governmental

Table 2.3. Categorization of Waste Management Strategies

Action	Agent	
	Government	Industry
Voluntary	1	3
	5	
Compelled	2	4

policies may seek to encourage voluntary behavior (cell 1) or may compel behavior through legislative mandate (cell 2). Industry initiatives may potentially include voluntary programs (cell 3) or more compulsory industry programs of self-policing (cell 4).

Within these categories, additional distinctions can come into play. For example, governmental strategies may be described as supply-side or demand-side as to the intended effect. Increases in the rate of source separation, mandatory deposits, material bans, and product requirements are examples of supply-side efforts. Demand-side tactics to increase demand for recycled materials include tax incentives and credits to industry for equipment, legislation to decrease the cost advantage of virgin materials (for example, HR 3737 would have set a $7.50-per-ton fee on the use of virgin materials including plastics, paper, batteries, and packaging), and governmental purchasing policies. In a similar fashion, voluntary industry responses may be further delimited either as efforts in support of public policy goals or as efforts to avoid or resist public policy goals.

The cells of Table 2.3 represent positions on continuous dimensions, rather than mutually exclusive categories established by rigidly dichotomous variables. The notion of movement along the two dimensions of this table permits us to focus on the middle area shown as cell 5. This cell represents the realm of possibilities for creative and strategic interaction between the politically fueled strategies of government (the public sector) and the economically fueled strategies of firms and entire industries (the private sector). That is, there is a verdant political economy of solid waste management characterized by diverse patterns of interaction between the public and private sectors.

Governmental Action. Governmental policies to reduce the amount of postconsumer waste entering the MWS include those that encourage voluntary action (cell 1) and those that mandate compulsory action (cell 2). Traditional public policies run a gamut that includes education, litter laws, source separation programs (mandatory and voluntary), recycling goals (mandatory and voluntary), requirements for reusable containers, zoning, and container deposits and fees. In 1986, New Jersey became the first state to establish a minimum mandatory goal of 25 percent reduction in residential, commercial, and institutional solid waste (cell 2; see Watson, 1989; National Solid Waste Management Association, 1989, pp. 1–2). Newer strategies include provision of technical assistance to industry to promote recycling, grants and loans to government and industry, product bans, material requirements, government tax credits to private recyclers for equipment, sales tax exemptions for recycling equipment and facilities, and grants to local governments to establish new programs of waste reduction. At least ten states and six local governments have enacted some form of content legislation. Florida, Connecticut, and California have enacted material content legislation for newsprint. Florida imposes a ten-cents-per-ton fee on newsprint, which will increase to fifty-cents per ton if a goal of 50 percent recycling is not met by October 1992. Connecticut requires newspapers with circulations of more than 40,000 to use at least 20 percent recycled materials. California requires that 25 percent of the newsprint consumed in 1991 be composed of at least 40 percent recycled newsprint.

Industry Action. Industry tends to favor government-sponsored collection, separation, and storage of waste materials. Industry generally has ignored voluntary government programs or adopted a proactive stance to avert legislation and avoid active participation in waste reduction. A number of public service and information organizations have been voluntarily created and supported by industry (cell 3). "Voluntary programs such as BIRP [Beverage Industry Recycling Program], Keep America Beautiful and Operation Curbside have apparently helped to quell the nation's political leaders" (Davis, 1986, p. 52). Some programs, such as Keep America Beautiful, have gained public acceptance. Others, such as the Council for Solid Waste Solutions (funded with $9 million the first year from plastics manufacturers; Matlack, 1989, p. 2401) and the Plastics Recycling Institute, are not as well known.

Voluntary recycling and litter clean-up programs have not entirely abated public pressure for mandatory deposits and material bans. In response to the adoption of deposits and material requirements, the packaging industry has been forced to initiate specific recycling programs (Lindorff, 1990). Pending passage of mandatory deposit legislation in Massachusetts, twelve beer wholesalers created their own recycling company, CRInc. CRInc originally served only the twelve wholesalers that owned the firm; the firm now services soft drink bottlers, distributors, and supermarket chains and recycles polyethylene (PET) bottles as well as glass (Galvin, 1988c).

Industry opposition to mandatory recycling produced BIRP as an alternative. BIRP is essentially a voluntary industry "buy-back" program for materials that are recyclable. However, there is no incentive to the consumer to return the container other than a payment based on the market value of the material. While the free-market approach is effective for aluminum cans, which have a relatively high price per pound, cash values for glass and plastic are considerably lower. Resource recovery plants, expensive to build and operate, are proposed as alternatives to governmental policies that compel industry to actively participate in waste reduction. Mandatory deposits are attacked as representing a piecemeal approach to the solid waste problem, which resource recovery plants address as a whole. Industry incurs no costs with this alternative.

In general, industrial organizations can also undertake stricter policies of self-regulation that direct member firms or practitioners to comply with certain requirements or procedures (cell 4). For example, professional organizations in fields such as medicine undertake functions of self-policing and certification. There are no clear examples of such industrywide policies, though there is clearly informal pressure for firms in some industry groups to join or contribute to lobbies that oppose legislation unfavorable to the industry. Governmental mandates are generally opposed by the affected industry, quite often effectively. The soft drink and beer industry has several well known and powerful lobbies, including the National Soft Drink Association, Glass Packaging Institute, Can Manufacturers Institute, American Iron & Steel Institute, National Beer Wholesalers of America, National Beverage Packaging Association, Wine and Spirits Wholesalers Association, National Food Processors Association, and the Society of the Plastics Industry. Not all of these groups have political action committees (PACs) to represent their interests. The following is a partial list of industry PACs with an interest in mandatory-deposit legislation and their 1985–86 expenditures/contributions: Aluminum Company of America ($71,924/$72,650), Anheuser-Busch Companies ($96,060/$93,310), IC Industries with subsidiary Pepsi-Cola ($72,060/$48,175), Owens-Illinois ($111,463/$105,600), PepsiCo ($120,597/$93,975), and the National Beer Wholesalers Association ($451,208/$362,350).

Public-Private Partnerships. In the mix of strategies that fit roughly within cells 1–4 of Table 2.3, it is increasingly possible to detect an emergent set of outcomes that synthesize what goes on in multiple individual cells. These emergent results are sufficiently novel and different from the strategies already discussed to be placed in a separate category, cell 5. In Table 2.3, cell 5 is overlaid on cells 1–4 in order to emphasize its synthesizing character. Outcomes in cell 5 emphasize negotiation and the interaction of initiatives from both public and private sectors; they can usefully be viewed as public-private partnerships. The partnerships that make up the contents of cell 5 are of two types: preemptive and compelled. The plastics packaging industry offers several examples of preemptive partnerships. Until recently, the plastics industry did not assume much responsibility for waste reduction (Haas, 1987, p. 67). But packaging control and material ban legislation that would affect the plastics industry has been offered in several states, including Connecticut, California, Illinois, Michigan, New Jersey, and Massachusetts (Prince, 1989). Several local governments have also proposed or adopted plastic bans, including New Haven, Milford, Hamden, and Guilford in Connecticut, and Minneapolis, Minnesota.

In reaction to the threat that restrictive legislation *could be* adopted, the plastics industry has instituted several recycling programs to preempt possible legislation, becoming, in effect, a partner with government. The National Polystyrene Recycling Company (NPRC) was recently formed by eight plastics manufacturers to recycle the plastic used in fast-food restaurants, including disposable cups and utensils; five regional recycling facilities are planned. The companies include Arco Chemical, Amoco Chemical, Chevron Chemical, Dow Chemical, Fina Oil & Chemical, Mobil Oil, Huntsmal Chemical, and Polysar Inc. McDonald's Corporation provides another example of preemptive partnership. McDonald's recently started a polystyrene recycling program for in-store customers in New England with the intent to expand the program to all of the company's restaurants in New England. McDonald's proposed to send its polystyrene to a recycling facility operated by NPRC. This effort was short-lived, however, as McDonald's announced the phase-out of its plastic foam hamburger container, an about-face from a company that steadfastly has insisted on the need to use plastic packaging. Finally, Proctor & Gamble Manufacturing Company, the leading manufacturer of disposable diapers, recently announced plans to develop an experimental recycling program to collect diapers and convert them to useable products. Industry appears willing in some instances to take preemptive action to avoid possible adoption of restrictive governmental policies, but the long-term impact of preemptive action must be questioned. Some industry observers question the industry's commitment to recycling, calling efforts such as these a "smoke screen set up by container and material suppliers to quiet legislators and environmentalists" (Davis, 1986, p. 52).

Industry may also be compelled to participate in a partnership under the threat of delayed implementation of legislation (cell 5). In the case of solid waste management, the compelled partnership revolves around the creation of industry incentives to reuse and recycle through the delayed implementation of restrictive governmental measures unless some conditions specified by government are met by industry. Industry may be compelled to act under the threat of governmental restrictions, such as mandatory deposits or material bans, and increase the rate of recycling to prevent implementation of more onerous policies; thus industry is compelled to adopt public goals. The case of recycling policy in Florida provides an example.

Compelled Partnership Strategy in Action: The Case of Florida

Recent legislation in the state of Florida provided a comprehensive program to deal with solid and hazardous waste management. Integral to this legislation was the use of mandatory deposits or advance disposal fees on packaging.

Florida's Solid Waste Policy. The Florida Solid Waste Act of 1988 (Chapter 88-130 Florida Laws) is a comprehensive approach to solid waste management. The act sets state recycling goals, requires local government solid waste programs, and provides annual and one-time grants for research in areas such as recycling, used oil, waste tires, and technology development. Grants in the amount of $20 million per year are distributed to local governments (over 50,000 population) to establish recycling programs. In addition, the act establishes tax exemptions for recycling equipment, bans cans with removable pop-tops, and bans ring connectors and plastics made with chlorinated fluorohydrocarbons. Plastic bags and take-out food containers must be biodegradable.

Of particular interest is the manner in which the act implements mandatory deposits. Advance disposal fees (deposits) are scheduled to take effect in 1992 on all containers that fail to attain a 50 percent recycling rate by that date. All containers made from aluminum, metal, glass, plastic, and plastic-coated paper are affected. Advance disposal fees will increase to two cents for containers that do not meet the recycling goal after 1994. This provision is a single-level approach to deposits and does not discriminate against the beverage industry.

The Florida act and legislative recommendations have overcome many industry objections to mandatory deposits and other forms of compulsory governmental policies. To understand why the Florida Soft Drink Association supported the Florida act, it is important to understand the political economy of financial incentives to control solid waste.

Political Economy of Financial Incentives. Deposit fee legislation is a vigorously contested issue, pitting the beverage and bottling industries against environmental advocates. Several federal studies attest to the effectiveness of mandatory deposits to reduce litter and solid waste, and to reduce the amount of energy and raw materials used in production (U.S. General Accounting Office, 1977, 1980; U.S. Congress-House, 1989). However, as one study reports, "These positive effects are purchased primarily at the expense of the beverage companies" (U.S. General Accounting Office, 1980, p. ii). The primary question is how do we place the burden of packaging disposal on the industry that produces these materials, and the consumer who uses them, with a minimum of governmental intervention?

Opponents of mandatory-deposit legislation argue that such laws will increase unemployment and are not effective in reducing litter, and that deposit fees single out the beverage industry, which contributes only a small percentage to the solid waste stream. These arguments have political appeal. However, they hide the real economic concern of the industry: the loss of market and the loss of profit generated through deposit fees. There are several broad factors that explain the shift from returnable to nonreturnable packaging: consumer reaction to deposits, consolidation of soft drink and beer bottlers into large-scale plants that depend on regional distribution systems, regional distribution systems require lightweight packaging, decrease in relative price advantage of returnable bottle over nonreturnable containers, and self-interest of glass, aluminum, and plastic manufacturers. Other industry concerns with mandatory-deposit legislation are more narrowly defined: transshipping and escheats, legislation generally targets only beverage containers and, most frequently, glass bottles, and tiered versus single-level approach to deposits.

First, industry is concerned with the reaction of consumers to perceived higher prices resulting from the imposition of container deposits. The beverage market is highly elastic: When the price goes up, consumption declines. When a deposit fee is enacted, the consumer frequently perceives it simply as a price increase. According to industry spokesman David Karmol, "In every state where there's a mandatory [deposit] law, the overall consumption drops, and it doesn't catch up for five or six years" (Davis, 1986, p. 52).

Second, there is the problem of maintaining the economies of scale created by consolidation. The beverage industry has experienced significant centralization over the past thirty years: "The growth of throwaway beverage containers over the past twenty years has contributed to a sharp decline in the number of soft drink plants. Larger bottlers benefit from the throwaway system because they can ship non-returnable soft drinks (particularly those in cans) farther than they can in refillable bottles. The result is that the larger metropolitan bottlers with their greater economic leverage are taking over smaller bottlers in outlying areas. In 1960, there were 4,519 bottlers in the United States. By 1970, the number had been reduced to 3,054. In 1976, the total had fallen to 2,246" (Norton, 1978, p. 398). The concentration of industry has resulted in just a few large companies, most of which are committed to the throwaway container. "This same process took place in the beer industry 20 to 25 years ago. In 1958, there were 262 breweries in the United States. By 1966, this had dwindled to 115, and [by 1978] there [were] fewer than 60 brewing companies left. Industry observers predict that there may be only 30 brewers left in operation within five years" (Norton, 1978, p. 398).

Third, the switch to nonreturnables has allowed major beverage companies to enter remote markets once closed by distance. Centralization, coupled with high-speed production, creates greater economies of scale. "It would take only a 40 percent increase in shipping radius to offset a 50 percent decline in the number of plants, and such an increase in shipping radius is indeed a plausible consequence of the switch from predominantly returnables to predominantly nonreturnables" (Landes and Posner, 1978, p. 198).

The factors of centralization, regional distribution, and high-speed production have generated increased profits for the industry. However, this restructuring has produced some negative externalities ignored by arguments against deposit legislation: Smaller bottlers have been forced out of business. Deposit legislation would have helped smaller producers by reducing their bottling costs and placing them on a more equitable footing in the area of distribution.

The introduction of PET bottles accelerated the demise of returnable bottles because they are lighter to ship. The first commercial PET bottles were produced in 1977 after Dupont issued its first license to manufacture PET bottles to Amoco Container Corporation in 1976. While PET bottles and containers are recyclable (they can be turned into other products), they are not reusable; they cannot be sanitized without remelting. Not all countries embrace PET bottles; Germany has banned the sale of bottles made from PET (Glazer, 1987).

Fourth, the soft drink industry has not shifted entirely to nonreturnable PET bottles for several reasons. Perhaps most important is that manufacturers and bottlers seek to promote a broader product base by offering their products in an increasing variety of sizes and container types. Each size and container type demands additional shelf space in the retail store. The second reason is that economics demand that industry evaluate the relative cost of production using aluminum, glass, and PET. Finally, the industry has expressed some apprehension over the impact of possible recycling legislation, and some firms may choose to produce and use packaging based on recycling concerns (Lang, 1989).

Finally, the glass-manufacturing industry is concerned with selective imposition of deposits on beverage containers and the potential impact of this targeting on the glass beverage container industry and market. Deposits have traditionally been levied against glass bottles alone. If every glass bottle were reused only once, a tripage rate of 1, the glass bottle market could conceivably be cut in half. A tripage rate of 20, not uncommon in the 1940s and 1950s, would have serious impact on the glass bottle industry and suppliers. Deposits on glass bottles would also reduce product diversification, which would affect the industry as a whole. This discussion is continued below in the discussion on tiered and single-level approaches to deposits.

More narrow issues of concern to industry include transshipping and escheats. Transshipping occurs when a distributor ships beer from a non-deposit state to a retailer in a deposit state. "Mandatory deposits encourage the transshipper to sell out of his territory cheaper by not charging the deposit value" (Calem, 1986b, p. 75). Retail stores selling transshipped products can offer them at a lower price because there was no deposit paid to the distributor. Consumers return the empties to various places, and eventually the wholesaler pays the deposit refund; the wholesaler receives more cans and bottles than are sold and is stuck with paying the extra deposits. The solution is brewery initiation of deposits (Galvin, 1988b).

The ownership of escheats (unclaimed deposits) is a contested issue. Only Michigan has succeeded in passing legislation to permit the state to collect unclaimed bottle deposits; in other states the unclaimed deposits are retained by the bottlers.

Another industry objection is that deposit legislation is selectively directed at the beverage industry. Opponents cite government data that show that beverage containers constitute only a small proportion of the solid waste stream (approximately 6 percent) (U.S. EPA, 1988). If half of all beverage containers were recycled, this would reduce the waste stream by only 3 percent. Opponents argue that the cost of compliance with deposit legislation for the beer and soft drink industry far outweighs the benefits that may accrue to the public.

The industry also objects when deposits are levied in a manner that favors one bottling material over another. Deposits may be levied in either a tiered or single-level approach. In a tiered system a single bottle material, generally glass, carries the mandatory-deposit fee. In a single-level system of deposits, all beverage containers carry deposit fees regardless of material. There are several impacts that result from the tiered approach. There may be a shift by bottlers from glass to aluminum and PET, which do not carry deposits. Retailers may reduce shelf space for glass bottles, preferring to focus on packages that are easier to handle. Some industry observers report that retailers may even eliminate glass beverage bottles to avoid handling deposits (Calem, 1986a, p. 45).

Dynamics of the Compelled Partnership Strategy. The Florida legislation creates an interesting and somewhat subtle situation for industry. The policy manages to avoid most industry objections to mandatory deposits (except for transshipping, a problem that could be resolved by national legislation requiring bottlers to accept any of their bottles from any source). Compliance is, in a strict sense, mandatory: The state is unambiguously exerting its coercive authority. A key aspect of the Florida policy, however, is that implementation of mandatory deposits will occur only if industry fails to meet recycling goals. The Florida strategy thus strongly encourages the development of feasible recycling programs by industry. The policy thus has the effect of creating a compelled partnership between industry and government. The legislation does this through a shift from imposing deposit fees on beverage containers for litter control, to imposing fees on all containers to manage solid waste. The legislation's reliance on voluntary efforts by industry, in addition to deposits on all containers without regard to material or use, has had the effect of creating industry support for the legislation. The Florida legislation demonstrates a manner, or configuration, through which industry and government can develop partnerships to share solid waste responsibilities, even if in a context of compulsion.

Conclusion

Solid waste generation will continue to grow, increasing disposal problems for states and local governments. While industry prefers to leave the solution to landfills and incinerators, there is increased resistance to these facilities as (1) construction costs escalate, (2) new requirements impose additional costs on operations, and (3) public opposition to new facility sites continues to expand (Spindler, 1989). Additional approaches to managing the solid waste dilemma are clearly necessary to develop a reasonably complete solution.

Preemptive and Compelled Partnerships. A new type of privatization strategy is evolving that involves government and industry in an innovative partnership with two variations: the preemptive partnership and the compelled partnership (cell 5 in Table 2.3). In both cases, industry acts to avoid restrictive legislation. In a preemptive partnership industry is reacting to legislation that has been threatened or proposed but is not yet adopted. In a preemptive partnership industry creates a program designed to reduce the demand for restrictive legislation. Without legislatively mandated goals, however, industry is relatively free to take the minimal action possible, which will reduce the threat of legislation. The preemptive partnership tends to favor industry over government.

In a compelled partnership industry is forced to react to legislation that will take effect unless specific actions are undertaken, or a mandated goal is met. This is sunset legislation in reverse, "sunrise" legislation, if you will. The legislation is adopted but delayed in implementation. The adoption or *perceived threat of adoption* of deposit fees and material bans or requirements by state and local governments has successfully compelled industry to initiate specific programs (Lindorff, 1990, p. 40).

Industry views compulsory government policies from the perspective of cost, market share, and profit. Industry has successfully passed the cost of solid waste disposal to government and the public at large. But this arrangement between government and industry is slowly changing. Some state and local governments are acting to require industry to undertake mandatory recycling at industry expense; they place the cost on the waste generator and on the direct consumers of the product, not on the general public.

Nonrecyclable packaging is overproduced because the costs of disposal (negative externalities), rather than being internalized in the price of the product, spill over into the public sector. By requiring product packing to be recyclable, and manufacturers to internalize part of the cost of disposal, government can reduce the negative externalities represented by the generation of solid waste. Success in such an effort requires careful examination of financial incentives.

Design of a Financial Incentive System. There are multiple options available and issues involved with regard to minimizing industry objections and maximizing the impact of financial incentives. Voluntary industry recycling can be encouraged through the use of tax credits and incentives; in this way the recycling infrastructure could be developed. Compelled partnerships can be developed with the adoption of mandatory recycling goals, supported with deposit fee legislation implemented in the event that recycling goals are not met.

Deposits should apply to all beverage containers regardless of materials, and, ideally, all packaging would be covered, not just beverage containers. Under this approach, industry would have the opportunity to demonstrate that voluntary recycling can work, and industry would have time to institute new programs before mandatory deposits were implemented. The compelled partnership approach provides a safeguard for the public interest if voluntary efforts are insufficient with the phased implementation of mandatory-deposit fees.

An additional benefit derived from the use of compelled partnerships is that the debate over the effectiveness of mandatory deposits can be closed. Termination of this debate will not be accomplished with a study that conclusively proves the effectiveness of deposit fees; rather, industry must be offered the opportunity to demonstrate that its "voluntary" efforts are effective, and its efforts must be supported through tax incentives to recycle and with a single-level approach to deposits. The lack of adequate data, and inadequate analysis of existing data, places the decision to use financial incentives in the political arena. That industry has lobbied so long and hard against mandatory deposits reduces the possibility that technical analysis could substitute for politics in any event.

Need for Additional Research. Determination of the composition of the MWS and evaluation of the impact of deposit fees on waste generation constitute little more than a fishing expedition at present. There is almost a singular lack of objective, scientific data on national MWS composition (Rathje, 1989). Some studies have attempted to extrapolate national waste-generation data from the waste collected in ten or twenty cities. The "materials-flow" method, typified by the Franklin Institute (U.S. EPA, 1986, 1988), is based on national industrial production figures and assumptions of disposal rates. These assumptions remain largely untested, yet the results form the basis for public policy debate. The lack of accurate data is a technical problem affecting the evaluation of waste stream reduction and management techniques. The lack of waste stream composition data makes the evaluation of the effects of mandatory deposits on the waste stream problematic. In conclusion, it is hoped that this research may offer some guidance to the states and Congress. Legislation that sets a recycling target and then imposes mandatory fees if the target is not met will likely encourage more industry action and commitment to recycling than will merely the threat of legislation not yet adopted.

References

Calem, M. "Learning to Cope with Bottle Bills." *Beverage World,* 1986a, *105* (1361), 44–48.

Calem, M. "Learning to Cope with Deposit Legislation." *Beverage World*, 1986b, *105* (1363), 72-76.

Davis, T. "Beyond the Point of No Return." *Beverage World*, 1986, *105* (1369), 52-55.

Galvin, A. "Glass Rebounds." *Beverage World*, 1988a, *107* (1418), 33-34, 43, 108.

Galvin, A. "The N.Y. Beer Mess." *Beverage World*, 1988b, *107* (1424), 85-90.

Galvin, A. "People Who Don't Recycle Plastic Are Crazy." *Beverage World*, 1988c, *107* (1410), 45-46.

Galvin, A. "Solid Waste Challenge." *Beverage World*, 1989, *108* (1443), 23-26.

Glazer, S. "Garbage Crisis." Editorial Research Reports, Congressional Quarterly, 1987, 2 (10), 474-486.

Haas, D. O. "PET Challenge." *Beverage World*, 1987, *106* (1393), 62-68.

Jeffords, J. M. Statement Before the Subcommittee on Transportation and Commerce of the House Committee on Interstate and Foreign Commerce. *Hearings on Mandatory Deposits on Beverage Containers*. 95th Cong., 2nd sess. Serial 95-184, 1978.

Landes, W. M., and Posner, R. A. Statement on Behalf of the National Soft Drink Association Before the Subcommittee for Consumers of the House Committee on Commerce, Science, and Transportation. *Hearings on S. 276, Beverage Container Reuse and Recycling Act of 1977*. 95th Cong., 2nd sess. Serial 95-102, 1978.

Lang, N. A. "Hoisting the Glass." *Beverage World*, 1989, *108* (1433), 36-40.

Lindorff, D. "The Paper Chase." *Governing*, 1990, 3 (6), 34-40.

Matlack, C. "Recycling Bandwagon." *National Journal*, Sept. 1989, pp. 2399-2402.

National Soft Drink Association (NSDA). *Forced Deposit Laws: There Are No Winners*. Washington, D.C.: NSDA, 1989.

National Solid Waste Management Association (NSWMA). *Recycling in the States: Update 1989*. Washington, D.C.: NSWMA, 1989.

Norton, N. E. Statement Before the Subcommittee for Consumers of the House Committee on Commerce, Science, and Transportation. *Hearings on S. 276, Beverage Container Reuse and Recycling Act of 1977*. 95th Cong., 2nd sess. Serial 95-102, 1978.

Organization for Economic Co-Operation and Development (OECD). *Beverage Containers: Re-Use or Recycling*. Paris: OECD, 1978.

Prince, G. "Minnesota Thunder." *Beverage World*, 1989, *108* (1433), 28-31, 88-89.

Rathje, W. L. "The Three Faces of Garbage—Measurements, Perceptions, Behaviors." *Journal of Resource Management and Technology*, 1989, *17* (2), 61-65.

Spindler, C. J. *The Effects of Commercial Products Packaging on the Management of Solid Waste in Florida*. Gainesville: Center for Solid and Hazardous Waste Management, State University of Florida, 1989.

U.S. Congress. House. Office of Technology Assessment. *Facing America's Trash: What Next for Municipal Solid Waste?* Washington, D.C.: Government Printing Office, 1989.

U.S. Congress. Senate. Subcommittee for Consumers of the House Committee on Commerce, Science, and Transportation. *Hearings on S. 276, Beverage Container Reuse and Recycling Act of 1977*. 95th Cong., 2nd sess. Serial 95-102, 1978.

U.S. Environmental Protection Agency (EPA). *Proceedings of the First National Conference on Packaging Wastes*. Washington, D.C.: EPA, 1971.

U.S. Environmental Protection Agency. *Characterization of Municipal Solid Waste in the United States, 1960 to 2000*. Franklin Report. Washington, D.C.: EPA, 1986.

U.S. Environmental Protection Agency. *Characterization of Municipal Solid Waste in the United States, 1960 to 2000*. Update of Franklin Report. Washington, D.C.: EPA, 1988.

U.S. General Accounting Office (GAO). *Potential Effects of a National Mandatory Deposit on Beverage Containers*. Series PAD-78-19. Washington, D.C.: GAO, 1977.

U.S. General Accounting Office. *States' Experience with Beverage Container Deposit Laws Shows Positive Benefits*. Series PAD-81-08. Washington, D.C.: GAO, 1980.
Watson, G. "Market Development Strategies for Recyclable Materials." *Journal of Resource Management and Technology*, 1989, 17 (2), 85–91.

Charles J. Spindler is assistant professor of political science and research coordinator, Center for Governmental Services, Auburn University, Auburn, Alabama. He has over thirteen years of public service experience, holding positions of city manager, director of community development, and community planner in a community action agency.

State regulatory agencies responsible for water quality typically have done little to develop policy or even provide information about privatization as a means for local authorities to develop wastewater treatment projects. Absent federal mandates, privatization takes place "in the shadow of positive law," and state agencies engage in it only informally and reactively, leaving private firms and local officials to carry out the privatization option.

How State Regulatory Agencies Address Privatization: The Case of Wastewater Treatment

John G. Heilman, Gerald W. Johnson

Major changes in national policy can generate major changes in the responsibilities of public agencies and officials for policy implementation and administration. So it was with the New Federalism in the 1980s. Two major principles of the New Federalism (Johnson and Heilman, 1987b) were and are devolution and privatization. Under the former, responsibility for the financing and administration of many policies and programs shifts from the national level of government to the state and local levels. Under the latter, responsibility shifts from the public sector to the private sector.

In practice, devolution and privatization are often interrelated. The connection tends to be especially visible when a driving force for both policies is national budgetary constraints. In other words, as the Gramm-Rudman deficit reduction process goes forward, we can expect to see a

This chapter was presented in an earlier version at the annual meetings of the American Political Science Association, Atlanta, Georgia, September 2, 1989. It draws on and completes the authors' paper "Privatization of Municipal Wastewater Treatment Works: Toward a Study of the View from the States," presented at the annual meetings of the Southern Political Science Association, Charlotte, North Carolina, November 8, 1987. The research reported here was supported by the U.S. Geological Survey, Department of the Interior, grant #14-08-0001-G1288, and by the Office of the Vice-President for Research and the Water Resources Research Institute, Auburn University. The contents of this chapter do not necessarily represent the policies of these agencies, and no assumption of endorsement by the U.S. government or any other agency should be made.

push in Congress to pass off to the states many burdens of program funding. As this takes place, there will be openings, even invitations, for responsible state agencies to allow or encourage the private sector to assist in the provision of services and of the capital needed to support them.

Thus, the manner in which state agencies address themselves to privatization options has potential bearing on public policy implementation in a range of areas. One area in which this interrelationship holds is municipal wastewater treatment. The present chapter explores how state agencies have addressed privatization in this service setting. It explains how the question arises in relation to national water policy, offers conceptual foundations for addressing the question, and provides some answers based on data from a national survey of state environmental regulatory agencies. The conclusions reached here have implications for the evaluation of many kinds of innovative local service programs over which state agencies exercise regulatory authority. That is, the conclusions apply to evaluation far beyond the ambit of sewage treatment, and even beyond the ambit of privatization.

National Clean Water Policy in Transition

The Clean Water Act of 1972 established both a national policy for the treatment of municipal wastewater and a national program for implementing that policy. The policy goal was to make rivers and streams fishable and swimmable. A specific objective was to clean municipal wastewater to appropriate effluent levels. The initial implementation mechanism was a program of nationally funded grants, administered by the U.S. Environmental Protection Agency (EPA) through state regulatory agencies. Though the legislation was enacted after four years of the Nixon presidency (for some of the political considerations involved in its passage, see Meier, 1985), its underlying premises were consistent with the philosophy of the Great Society: National problems were subject to nationally developed and nationally funded programs to be implemented by the federal government through state administrative agencies.

Although the overall impact of the Clean Water Act on the quality of the nation's water remains to be assessed (but see U.S. General Accounting Office, 1987), it is clear that project grant funding became a central mechanism through which municipal wastewater treatment works were designed and constructed. Since 1972, over $40 billion in grants have supported the development of more than one thousand municipal plants nationwide. The Water Quality Act of 1987 replaced federal grants with state loans as a means of project funding, but this change did not reduce the potential importance of privatization as a method of project development and a means for implementation of national water policy objectives.

Federal support of water projects over the past twenty years has proceeded through the institutional medium of state environmental regulatory agencies. Thirty-eight state environmental regulatory agencies received authority from EPA to administer clean water policy in their respective states. Their main regulatory activities include issuance of permits, review of plant design and construction, grant administration, monitoring of plant effluent and stream quality, and negotiation of compliance and sanctions.

After the passage of the Clean Water Act in 1972, federal funding for project grants flowed at a steady pace for the better part of a decade. Municipalities were able to obtain up to 75 percent or more of project costs in the form of federal grant support. In the early 1980s, however, the situation changed dramatically. In line with President Reagan's New Federalism (Johnson and Heilman, 1987b), appropriations for grant funding were reduced, as was the percentage of project costs that could be funded through grants. Furthermore, EPA toughened up its enforcement policy. It took a firm position that municipalities must comply with national standards by a deadline of July 1, 1988, with or without federal funding. Thus, many cities were faced with a need to renovate existing facilities or to build new facilities at the same time grant funding was reduced.

As these changes in water policy took effect, Congress enacted tax code revisions that gave the private sector incentives to invest in and develop public works projects. In the wake of these revisions, private engineering and financial firms developed a comprehensive privatization strategy through which a city could engage a single service provider to undertake the financing, design, construction, ownership, and operation of a treatment work (Goldman and Mokuvos, 1984). The comprehensive privatization approach offered a potential savings in time and money to cities eager to proceed with project development.

On the basis of this savings potential and the tax benefits that became available in 1981 and 1982, privatizers were able to offer deals for plant design, construction, and operation at a total cost well below the cost of other options. Throughout the mid-1980s, many cities across the United States explored this comprehensive privatization option, and several proceeded with it (Heilman and Johnson, in press). It should be pointed out that although tax law changes in 1986 slowed a mounting wave of privatization to a trickle, this particular strategy continues to be of interest in a number of fields. The private sector has lobbied Congress since 1987 in favor of restoring the tax incentives eliminated in the general tax reform of 1986.

The first cities that considered the option of privatizing faced numerous barriers. Some of these barriers had to do with state laws that prohibited several actions that privatization required. For instance, cities needed the authority to sign long-term contracts, issue special tax-exempt bonds, and let bids on a noncompetitive basis. In several cases, cities wanting to privatize had to work with their state legislatures to achieve passage of legislation authorizing such actions.

Beyond the legal barriers, cities faced much complexity and uncertainty. The comprehensive privatization strategy was extremely complicated: In one early deal a municipality entered into 106 separate contracts. Furthermore, the comprehensive approach was an innovation in the wastewater treatment field. Thus, it presented the players, including the privatizing municipality, with multiple unknowns. For instance, how would privatized deals work out over the twenty or twenty-five years that contracts usually covered? What would happen if the privatizing firm went bankrupt, or decided to sell the facility to another private firm, or failed to operate the facility up to the treatment levels required by national law?

At the federal level, EPA offered no answers to these questions. The provision of regulatory advice and guidance to municipalities thus fell to state environmental agencies. This role was appropriate for them, since they in any case would exercise authority over a treatment plant, regardless of how it had been financed. In the case of grant-funded projects, state agencies exercised considerable and well-defined control over all aspects of treatment plant design, construction, and operation. They not only provided grant financing but also reviewed and approved project designs, monitored construction, and monitored pollution levels in plant effluent on a permanent basis. Especially during design and construction, the grant process gave state agencies a highly effective instrument of regulatory control: They could withhold money to compel compliance with their standards.

By contrast, the role of state agencies in privatization cases was unclear. Since privatized projects by law could not use federal grant money, agencies lost an effective mechanism of direct control over project funding. That is, the functions of monitoring and approving plant design and construction were unhitched from the function of ensuring accountability in the expenditure of federal and state money. And although state agencies retained responsibility for monitoring plant discharge and enforcing federal standards, it was not clear whether they should address the privatizing firm or the municipality to compel compliance with effluent standards.

This uncertainty notwithstanding, the early cases of privatization made clear that state environmental regulatory policy, or the lack thereof, would inevitably shape the development, adoption, and implementation of this innovation. First, these agencies represented an obvious potential and authoritative source of information and advice for local authorities. Second, they remained responsible for monitoring and enforcing plant compliance with federal affluent standards. And, third, state agencies represented the principal organizational framework available for actively insuring public accountability in privatized projects. They were far more directly involved in these cases than was EPA, which studiously avoided taking any public position on privatization. And state agencies were far more important as agents of accountability than courts could be. The role of the courts was potential rather than actual, and they would be engaged only in cases of formal legal dispute, not in the political negotiations that form a routine part of project development and ongoing operation.

Over a period of four years, from 1983 to 1986, the level of privatization-related activity rose dramatically across the country. Engineering and financial firms developed expertise in this field and advertised it in training seminars and literature aimed at local officials. A growing number of municipalities considered the privatization option. During this period, for the reasons set forth above, state environmental regulatory agencies were uniquely positioned to shape the development and adoption—the diffusion—of privatization as an innovation in water policy implementation.

How state agencies chose to address, or not to address, this new strategy was crucial to the impact that they would have on its use, and indeed on its success. Since privatization in the wastewater treatment field was an innovation and worked differently than in other policy areas where it had already been used, state agency officials had not had any opportunity to engage it. They did not know or had not thought about what it was, how it worked, what its advantages and pitfalls might be, who was able to do it, what was needed to make it happen, how it related to other possible financing options, how it related to state law, or what strategies and policies they might develop to regulate it.

In other words, at the point privatization was introduced, in about 1983, state agencies were a blank slate with respect to it. In the years that followed, however, officials in those agencies had the opportunity to address privatization: to develop information, perspectives, and positions with respect to it. The central question of interest in this chapter is how agencies proceeded in this regard. Did state officials become informed about privatization, and, if so, how? What did they know, think, and tell city officials about it? Did they address the relationship between privatization and other options for project development and financing? Did they make evaluations and form preferences with respect to privatization? Did agencies develop an informal position on privatization, or review its status as an option under state law? Did they develop formal policies with respect to it?

These questions are of interest in policy areas that extend far beyond wastewater treatment. In an era of increasing state-level responsibility for policy implementation and financing, such questions are likely to arise repeatedly across a broad spectrum of public service programs. Where they do so, answers to them can be of interest and use to evaluators charged with assessing privatization strategies. When implementors of new service configurations (Wise, 1990) must deal with external organizations in complex ways—as local officials had to deal with state environmental agencies—then the manner in which those organizations address the new arrangements (such as privatization) can powerfully influence program operations and impact. In this chapter, we address these matters through both theory and data.

State Regulatory Agencies as Agents of Change: Theoretical Perspectives and the Need for Data

The preceding section suggests a central role for state environmental regulatory agencies in the development of privatization as an implementation mechanism. To put it differently, state agencies will influence the emerging "design" of the privatization option. As the growing literature on policy design points out (Ingraham, 1987), one principal ingredient in this process is systematic information.

On a closely related issue, Savage (1985, pp. 15–16) states that among change agents, "only the federal government has received any systematic attention in comparative analyses of policy diffusion." He goes on to suggest that investigations are needed of the entire gamut of public, semipublic, and private organizations that plausibly can affect diffusion processes. (For a detailed analysis of the diffusion of wastewater treatment privatization, see Johnson and Heilman, 1987a). In other words, students of the processes by which innovative policies spread across the states should find it worthwhile to look at the role of state agencies and their interactions with a range of groups and organizations.

Broader reasons for such an investigation are offered by Feller and Menzel (1977), Thompson and Scicchitano (1985), Rogers (1983), and Carroll, Flynn, and Dorsey (1979). Thompson and Scicchitano address the question of what will happen to the vigor with which national policies are implemented as responsibility for these policies shifts to the states. By their account, the degree of energy that states bring to their tasks is an empirical question and requires comparative analysis across states. Feller and Menzel add that how states handle "new technologies" to increase their operating efficiency will be a function of "diffusion milieus," which will vary from state to state and can be examined empirically.

Rogers (1983, pp. 333–337) observes that many diffusion systems are more decentralized than is contemplated in the "classical" centralized model of, say, agricultural extension. Presaging the argument of Weiss (1988), he goes on to say that in the late 1970s he "gradually became aware of diffusion systems that did not operate at all like . . . relatively centralized diffusion systems. . . . Instead of coming out of formal R&D systems, innovations often bubbled up from the operational levels of a system, with the inventing done by certain users. The new ideas spread horizontally via peer networks, with a high degree of re-invention occurring as the innovations are modified by users to fit their particular conditions. Such decentralized diffusion systems are usually not run by a small set of technical experts. Instead, decision making in the diffusion system is widely shared with adopters making many decisions" (p. 334).

The situation of state agencies approaching implementation of state revolving loan funds (SRFs) and considering privatization appears to fit this description well. For our purposes, Rogers is suggesting that as the policies of the New Federalism are implemented, and as responsibilities increasingly devolve from the center to the periphery, state-level actions will more and more serve as a source of understanding of innovation processes. In sum, there is a need for studies of the processes by which state regulatory agencies participate in the development of innovative approaches to implementing national policy. Privatization is such an approach.

Beyond the practical need for information as a basis for policy design, state regulatory processes with respect to privatization are of interest for theoretical reasons. The following discussion develops theoretical under-pinnings for an examination of state regulatory processes in this regard.

In a broad-ranging study of regulation, Meier (1985) observes that as the New Federalism proceeds, and regulatory responsibility shifts corre-spondingly, new issues will arise. Some of these issues will be empirical. For example, in the case of comprehensive treatment plant privatization, an immediate question concerns the extent to which state agencies are engaging this option and have developed information, issues, points of view, and mechanisms for proceeding with it. And some theoretical issues will arise as well. For instance, when national mandates remain in place, but responsibility for accomplishing them is shifted to the states and national funding for implementation is eliminated, what kinds of responses can we expect from the relevant state agencies, and what factors will shape these responses? (For a review of federal incentives in policy diffusion and implementation, see Welch and Thompson, 1980.)

Some building blocks of theory to address these issues appear in works by McGarity (1986) and Treiber (1985). McGarity (1986, p. 416) advocates a "comprehensive analytical rationality" model of thought for agency personnel. In brief, this approach emphasizes regulatory attention to all aspects of and all potential solutions for a given problem. While this "model" is intended to describe how regulators might approach problems rather than how we might conceive of what regulators do, it appears rele-vant to the present issue. In particular, it suggests that to understand the regulatory process, we need to attend to the many ways in which regulators engage many possible solutions to problems, not just how they implement a particular solution.

Treiber (1985, p. 257) speaks of a "crisis of modern rationality" in the administrative state. The crisis is apparent, Treiber suggests, in the failure of welfare programs in the United States and environmental programs in West Germany in the 1970s. In his view, these failures make clear that "legal control," meaning regulatory outcomes, are not "expressed by condi-tional programming—by precisely defining factual conditions and legal consequences." In examining regulatory processes, Treiber urges that "the legal control of social action is indirect and abstract, for the legal system only determines the organization and procedural premises of future action." Certainly, this description appears to fit the case of clean water policy over the past two decades.

From this premise Treiber proceeds to describe the kinds of organizational behaviors that merit attention in evaluations of regulatory outcomes. He draws attention to informal, "reflexive" mechanisms, which he argues have long existed "in the shadow of codified law." He argues (Treiber, 1985, p. 257) that "formal law requires . . . informal structures if it is to be implemented at all in the face of the usual restrictions (imprecise and conflicting aims, problems of construing flexible legal terminology, low control and sanction potential, low motivation of those to whom norms are addressed)."

Treiber (1985, pp. 257, 261–262) goes on to suggest that we can learn much about the way in which agencies actually implement law by studying these informal, reflexive mechanisms. They include activities such as information gathering, discussions, nonbinding negotiations, tentative decisions and reviews of them, and exploratory contacts with interested groups inside and outside of the agencies. In Treiber's view, the nature of these additional groups—the constituents of the issue network to which Weiss (1988) refers—is particularly important. Treiber (1985, p. 258) sees the modern regulatory agency as operating in a condition of "semi-autonomy. . . . It can generate rules and customs and symbols internally, but . . . it is also vulnerable to rules and decisions and other forces emanating from the larger world by which it is surrounded.

These ideas suggest which kinds of regulatory behavior and which aspects of the regulatory setting deserve particular attention. They seem particularly appropriate to the present research area for more than one reason. First, even though Treiber's framework for analysis addresses regulatory politics in what was West Germany, it speaks to exactly the kind of fluid, transitional situation that exists as policy responsibilities devolve from federal to state agencies under the New Federalism. Second, Treiber's approach is well suited to examining regulatory processes as they relate to implementation options (for example, privatization), which are not clearly defined in a legislative mandate and are not routinely the subject of extensive and clear rules or guidance from the federal agency.

In sum, Treiber's contribution calls our attention to the informal processes by which agencies develop information, attitudes, and policies with respect to emergent issues and options. Where these issues and options have developed through private-sector initiatives rather than through positive law, the relationship of Treiber's ideas to those of McGarity becomes apparent. Their joint recommendation about the study of national policy implementation might read as follows: Examine the role of state agencies. What options are available for implementing national policy? Are some of these options more firmly grounded in positive law than others? How do the agencies engage these different options? What do they know, want to know, think, do to learn, tell others, and decide with respect to these options? How do these options interact with one another, and how do the agencies deal with the potential for interaction? What groups do the agencies communicate with on these issues, and what are their communications about? What formal decisions and positions do agencies establish, but also what kinds of informal, "reflexive" activities take place, or fail to take place, as implementation proceeds?

State Agencies and Privatization: Empirical Evidence

To answer some of these questions, and others as well, state environmental regulatory officials were surveyed in summer 1988. For purposes of the present study, the timing of data collection was ideal. The first round of privatization of wastewater treatment works was complete, and EPA had not yet promulgated policy in this area. Within a few months after data collection was completed, EPA held a conference in Washington, D.C., to herald a commitment to "public-private partnership." The data reported here represent the situation prior to this initiative by EPA.

Environmental regulatory agencies in the states were contacted by telephone to identify an appropriate respondent for a questionnaire for each state. That person was the agency official most likely to become directly involved with issues of privatization. Typically, it was the person in charge of compliance, permits, or construction grants under the existing system of EPA support. The person in this position therefore knew in virtually all cases how the state agency had addressed privatization.

While multiple interviews in each agency would have been helpful, they proved impractical for several reasons. First, in many cases only the person in charge was able, or willing, to address these matters with outside interviewers. Second, in many cases, personnel turnover was so frequent that only one person had information about the collective experience. Third, our experience in site visits indicated that there was variation across states in terms of the kinds of persons, in the interviewees' agencies or in other agencies, with whom informal interactions might occur. Fourth, because of routine scheduling conflicts and the time needed to collect information in response to some questions, the completion of the survey took a lot of time. Although the respondents seemed sincerely interested and helpful, many callbacks were required in some cases. Therefore, we limited our sample to one official per state.

The purpose of the survey was reviewed with the respondent, and a copy of the instrument was mailed. About a week later a follow-up telephone call was made to provide assistance as needed in completing the survey form. The responses were provided either by phone or through mail return of the completed instrument, depending on the preference of the respondent. The results here are based on the thirty-nine survey cases that were completed. We focus here on those results that relate to agency perceptions and the policy process. The eleven noncompletions resulted from a variety of factors, including heavy workload, recent departure of a key person who could have answered questions, and in some cases simple reluctance to proceed. The interviewers felt that the respondents who did complete the interview process did so with interest and sincerity, and in general there was no discernible systematic difference between the agencies that responded and those that did not.

In reviewing the results of the survey, we considered the following matters with respect to how states, as represented by their responsible regulatory officials, engaged privatization. First, were they aware of privatization, and did they regard it as feasible in general? (They were and they did.) Second, did state agencies make explicit connections between privatization and other implementation options? (They did not.) Third, did states do anything to address privatization in the terms discussed in the preceding section? (They did.) Fourth, was there variation in the extent to which states addressed privatization? (There was.) Fifth, are there detectable patterns in the variation observed? (There are, to a limited extent.) And sixth, what can we conclude from the data and analysis?

The survey results clearly indicated that state agency officials were aware of privatization. Thirty-four of thirty-nine (89 percent) reported that they had read about it in industry journals. They had additional knowledge. Seven respondents (18 percent) reported instances of privatization in their states, and officials in sixteen states (41 percent) indicated that cities or other authorities in their jurisdictions had seriously considered privatization.

The respondents were split, however, on the overall feasibility of the privatization option in their states. Nineteen (49 percent) said it would be feasible, 12 (31 percent) said it would not, and the rest were undecided. When asked what barriers might stand in the way of privatization, they also gave mixed responses. Table 3.1 summarizes information on the types of barriers perceived by the respondents. The question on political barriers was phrased in terms of the roles of the public and private sectors. The question on administrative barriers referred to accountability and responsibility of the agencies involved.

Additional information on the respondents' assessment of the private-sector role in water treatment emerged from a question on factors affecting water quality. Respondents ranked six factors "in terms of the amount of positive effect they had on improving wastewater treatment" in their state in the past five years. The answers provided added information on perceptions. Ranked in order of decreasing amount of positive effect, the factors were as follows: construction grants (most positive effect), permit process, clean water standards, regulatory sanctions, local desire for clean water, and private-sector activity (least positive effect).

Table 3.1. Barriers to Privatization Perceived by State Agency Personnel

Type of Barrier	Number of Respondents
Economic	16 (44%)
Political	13 (35%)
Legal	9 (25%)
Administrative	8 (22%)
Other	7 (21%)

Note: N = 39

In summary, state agency officials who were responsible for water policy implementation were aware of privatization and had "engaged" it to the extent of having some knowledge about its occurrence in their states and some opinions about it. While officials were divided on the feasibility of privatization, they were nearly unanimous in their evaluation of the overall role of the private sector in contributing to water quality. That is, they placed it at the bottom of the effectiveness pile.

Against this background, it is not unreasonable to ask our second question. Since officials knew about privatization, and, in several cases, cities in their states actively considered it, had their agencies developed a policy with respect to it? In this case the answer was also unambiguous. Thirty-four of thirty-nine respondents (87 percent) reported no policy toward privatization. In sum, there was awareness and some evaluation of this option, but very little in the way of direct engagement in the form of established policy.

This combination of findings leads to a pair of questions. Why were there so few formal determinations of policy? And, absent such formal determinations, what if anything did agency personnel do to address the privatization option? To respond to this third issue, we turn to the longest series of items in the survey. Respondents were asked whether they had engaged in fourteen different activities during the preceding year. The responses are shown in Table 3.2 ("it" refers to privatization).

The majority of respondents had read about privatization in trade journals, discussed it informally in the agency, and had discussed privatization with an industry agent. Formal experience with or actions taken regarding project privatization, including receipt of EPA guidance, were reported by less than a majority of respondents.

The results of the state agency survey support the findings based on on-site agency interviews and also cast new light on several issues. Perhaps the strongest pattern running through the data concerns the role of public agencies and private industry in the development of privatization. Industry initiatives clearly drove the privatization option. The state agency position was, at best, passive, not active. The two most extreme response distribu-

Table 3.2. Experience of State Agency Personnel with Privatization

Past-Year Experience	Number of Affirmative Respondents
Read about it in trade or industry journals	34 (87%)
Discussed it informally in agency	26 (67%)
Discussed it with industry or business agent	24 (62%)
Wanted more information about it	19 (49%)
Discussed it with municipal agent	18 (46%)
Received an inquiry about it	17 (44%)
Attended a seminar on it	14 (36%)
Received EPA guidance concerning it	11 (29%)
Assessed the feasibility of it under state law	11 (28%)
Wanted more EPA guidance	10 (26%)
Talked about it with state legislator or staff person	10 (26%)
Attended in-house meeting at which it was on the agenda	9 (23%)
Requested information about it	6 (15%)
Adopted an agency position (formal or informal) on it	5 (13%)

Note: N = 39

tions underscore this interpretation. Eighty-seven percent of respondents indicated that they had read about privatization in industry journals; the same percentage indicated that their agencies had taken no position on it, formally or even informally.

As the data just cited suggest, the pattern of policy processes in the state agencies related to privatization appears informal, internal, and reactive, rather than formal, external, and proactive. Respondents from 67 percent of the agencies indicated that they had had in-house discussions (thus, in fully one-third of the state agencies, privatization may not even have been discussed). The percentages were typically lower for behaviors that are more formal or outward looking: 23 percent reported having attended an in-house meeting, and only 13 percent had adopted even an informal agency position. Only 28 percent had taken a step toward policy in the form of evaluating privatization under state law, and just 26 percent had discussed it with a state legislator or staff person.

The pattern of discussions concerning privatization also is revealing. The primacy of industry activity appears again: 62 percent of respondents had discussed privatization with industry or business representatives. And the role of the municipalities as "partners" of the private sector is reflected in the 46 percent of respondents who had discussed privatization with municipal representatives.

Absent from the data is any sense of movement in the state agencies toward active engagement of issues concerning privatization. Although 44 percent of respondents had been asked about it, only 15 percent had made an inquiry about it. Barely half (49 percent) wanted more information about it, and just a quarter (26 percent) reported wanting more EPA guidance than they had received concerning it.

The responses evaluating factors in water quality improvement suggest that the focus of the state agencies is primarily on the regulatory process and its instruments: Grants, permits, standards, and sanctions were all rated higher than public interest in clean water. The role of the private sector ranked last by a wide margin: 82 percent of respondents placed it lowest.

Thus, it might be suggested that the view from the states focuses on public mandates rather than on private incentives as the more productive force for achieving clean water policy objectives. Indeed, it appears plausible that the state agencies view their mandates and procedures as the motor and transmission driving local authorities and the public toward clean water objectives. The notion of mandates suggests a theory-based explanation for the lack of organized initiative at the agency level to deal with privatization. Welch and Thompson (1980) concluded that the existence of federal incentives plays a central role in the diffusion and implementation of mechanisms for achieving national policy goals. Where incentives are lacking, the states are less likely to address the mechanisms in question. At the time these survey data were collected, privatization was a nonissue in terms of clean water policy implementation. No incentives, either positive or negative, existed for states to address it. By contrast, the grants program contained substantial positive (grants) and negative (fines) incentives. Thus, from the perspective of Welch and Thompson, it is hardly surprising that relatively few states undertook positive initiatives with respect to privatization. The municipalities that privatized water treatment facilities in the 1980s did so because of economic incentives, primarily pressures from industrial and residential developments (Johnson and Heilman, 1987b). State agencies were not confronted with this type of incentive directly.

There clearly was variation in the extent to which states addressed privatization either passively or actively. The responses in Table 3.2 show that some steps or events were much more frequent than were others. And a review of the responses on a state-by-state basis indicates that these steps or events were not evenly distributed across states. That is, some states reported that none of these things had happened, some reported that all of them had happened, and some reported a mix.

The variation in results leads to another question: What patterns, if any, can we detect in this variation? Here the results are mixed. One approach yielded, at best, sparse results. We assigned the states scores reflecting the number of events or steps that had taken place. The low score was zero (nothing had happened), and the high score was fourteen (they had seen or done it all). In this way it was possible to identify the extent to which, overall, individual states had "addressed" privatization.

This ordering of cases made it possible to assess several hypotheses. We thought, for instance, that states that had more personnel dealing with water policy might have done more to address privatization, since the larger staffs would quite possibly represent both more differentiated skills and interests and greater likelihood of slack resources to allow a staff person to deal with this new option. However, there was no relationship between staff size and intensity of privatization-related activity.

We also examined whether the variety of methods that a state used to assist in project funding was related to effort directed toward addressing privatization. The idea here was that if an agency had expertise in one or more ways of assisting municipalities to obtain funding (the options were state grants and state loans), it might be more disposed to examine yet one more alternative source of funding. Again there was no relationship.

Two additional variables proved unrelated to the states' overall activity with respect to privatization. First, states that had developed some policy with respect to SRF implementation were no more likely to have addressed privatization than were states that had not yet developed plans for SRF implementation. Perhaps more surprising, the matter of whether a city had actively considered privatization in the state was unrelated to the extent to which the state agency addressed privatization, as described through the questionnaire items.

Thus, our data suggest that it is not easy to determine what kinds of state agencies will actively address privatization. Again, we attribute this pattern—or rather, lack of pattern—to the absence of federal incentives, mandates, or even clearly defined policy of any kind with respect to privatization. In the concluding section of the paper we comment on the possible broader significance of these results. Before turning to the conclusions, however, we present one additional set of results that does point to the presence of an ordered and cumulative structure in what the states do when they address privatization. Here the question shifts from "What kinds of states address privatization actively?" to the more descriptive "When states address privatization, how do they do it?"

To answer this question we applied the Guttman scale technique to the responses of the thirty-nine state agency officials to the fourteen items shown in Table 3.2. The responses on eight of the items form a scale having a coefficient of reproducibility of .92. The results for these eight items are shown in Table 3.3 (see Miller, 1970, p. 93; Babbie, 1989, p. 409; on the use of Guttman scales).

Table 3.3 was constructed as follows. The state agency personnel were scored in terms of the total number of items (out of the eight) that they reported as having happened. Scores ranged from eight items (top of the list) to zero items (bottom of the list). The items were arranged from left to right in terms of how frequently they were reported. For instance, the left-most item (RE, read about privatization in trade or industry journals) was

Table 3.3. How State Agency Personnel Engage Privatization

State	RE (34)	IN (26)	BU (24)	CI (18)	LE (10)	MT (9)	QU (6)	PN (5)	N Yeses
IL	+	+	+	+	+	+	+	+	8
NJ	+	+	+	+	+	+	+	+	8
MN	+	+	+	+	+	+	+	+	8
VT	+	+	+	+	+	+	+	0	7
MD	+	+	+	+	+	0	+	0	6
DE	+	+	+	+	+	0	0	0	5
NM	+	+	+	+	+	0	0	0	5
CT	+	+	+	+	+	0	0	0	5
UT	+	+	+	+	+	0	0	0	5
LA	+	+	+	+	0	+	0	0	5
NY	+	+	+	+	0	+	0	0	5
OK	+	+	+	+	0	0	+	0	5
WI	+	+	+	0	0	+	0	+	5
MO	+	+	+	+	0	0	0	0	4
OH	+	+	+	+	0	0	0	0	4
KS	+	+	+	+	0	0	0	0	4
RI	+	+	+	0	0	0	0	0	3
AR	+	+	+	0	0	0	0	0	3
CO	+	+	+	0	0	0	0	0	3
TN	+	+	+	0	0	0	0	0	3
FL	+	+	+	0	0	0	0	0	3
WV	+	+	0	+	0	0	0	0	3
AK	+	+	0	+	0	0	0	0	3
HI	+	0	0	+	0	+	0	0	3
VA	+	0	+	0	0	+	0	0	3
MS	+	0	+	0	0	0	0	0	2
NE	+	0	0	0	0	0	0	+	2
AZ	+	0	+	0	0	0	0	0	2
MT	+	+	0	0	0	0	0	0	2
IN	+	+	0	0	0	0	0	0	2
ND	0	+	0	0	+	0	0	0	2
ID	+	0	0	0	0	0	0	0	1
OR	+	0	0	0	0	0	0	0	1
KY	+	0	0	0	0	0	0	0	1
SD	+	0	0	0	0	0	0	0	1
IO	0	0	0	0	0	0	0	0	0
CA	0	0	0	0	0	0	0	0	0
WY	0	0	0	0	0	0	0	0	0
ME	0	0	0	0	0	0	0	0	0
N Errors	4	3	3	6	1	5	1	2	

Note: Total $N = 39$; coefficient of reproducibility = 1 − (total errors/total responses) = 1 − (25 errors)/(8 items × 39 responses/item) = 1 − (25/312) = .92; + is "yes" and 0 is "no." The legend for the eight items is as follows: RE—read about privatization in trade or industry journals, IN—discussed it informally in agency, BU—discussed it with industry or business agent, CI—discussed it with municipal agent, LE—talked about it with state legislator or staff person, MT—attended in-house meeting at which it was on the agenda, QU—requested information about it, and PN—adopted an agency position (formal or informal) on it.

reported by thirty-four respondents, more frequently than for any other of the eight items.

To count errors, we added the number of "yeses" that appeared in each column below the topmost "no." A "yes" is shown by a plus sign (+), a "no" is shown by a zero (0). For instance, the column of responses for RE shows four "errors," since four "yeses" (+) appear below the first "no" (0). In this case, the first "no" came from ND (North Dakota).

Thus, although it was not clear which states would proceed to address privatization, a firm structure of activity characterized those states which did so. The steps typically consisted of reading about privatization in trade or industry journals, discussing it informally in-house, discussing it with an industry or business agent, discussing it with a municipal agent, discussing it with a representative of the state legislature, meeting formally about it in-house, then going out and asking questions about it, and, finally, reaching an agency position on it.

Care is needed in generalizing from these particular results. As Babbie (1989, p. 411) points out, "Scalability is a sample-dependent, empirical question." With this constraint in mind, the findings suggest that even where implementation options are not driven by federal mandates or incentives, state agencies that address those options do so in steps that reflect a common pattern. To the extent that this pattern obtains, it can assist in the assessment of policy options. As an example, EPA has considered establishing a national privatization clearinghouse as part of its recent initiative in public-private partnerships. The finding that state agencies seek information about privatization relatively late in the process of addressing it suggests that there is fertile ground within state agencies for clearinghouse information concerning privatization.

Also, more generally, it is of interest to find some consistency across state agencies in the manner in which they proceed to address a new option such as privatization. As responsibility for policy implementation and financing devolves upon the states, they will certainly have many occasions to consider novel approaches to achieving their objectives. Their ability to do so will depend in part on how they proceed. These findings suggest that there is a discernible structure to what they do. And, it relies heavily on the kinds of informal, reflexive mechanisms to which McGarity (1986) and Treiber (1985) call attention.

In this regard, it is noteworthy that all of the scalable items involved a form of communication rather than an attitude or an assessment. Also, six of the eight scalable items involve contact with a group or organization outside of the regulatory agency. The groups in question seem to be physically near at hand. They include industry, the state legislature, and municipalities. The absence of EPA from these items perhaps reinforces the local perspective that emerges from the broader set of responses of the state agency officials. One might cautiously suggest, then, that the state agencies reflected in these data are indeed functioning as independent laboratories for experimentation.

Conclusions

Based on the data and discussion of the survey presented here, several conclusions appear warranted.

1. As of mid-1988, responsible officials in state agencies were aware of privatization as an option and had had the opportunity to learn about and engage it to the extent that they wished to do so.

2. As reported by these same officials, the posture of the state agencies with respect to this option was typically informal, internal, and reactive, rather than formal, proactive, or externally oriented.

3. The data suggest an agency orientation toward public mandates and regulatory processes as keys to achieving clean water objectives, rather than an orientation toward citizen commitments or private-sector activity. This orientation has implications for privatization as set forth in the following.

4. The private sector, the municipalities, and other local authorities were carrying the privatization option.

5. Many state agencies appeared unready to engage the privatization option positively or to provide positive assistance with respect to it. The absence of federal incentives for the states to address privatization contributes substantially to this result.

6. The importance of state agencies in the development (or nondevelopment) of privatization options is increasing under the Water Quality Act of 1987. Thus, the prospects for privatization will depend in part on the extent to which state agencies are able—and enabled—to engage this option more directly and proactively than they have demonstrated previously. One step in this process could be the externalization or outward orientation and legitimation of this issue for the agencies. A mechanism to contribute to this process is already engaged: In October 1988 EPA formally acted to organize and promote public-private partnership structures as options for meeting national water objectives.

7. Although many states did not address privatization actively, some did. To the extent that they did so, they followed a fairly clear pattern of informal reviews and contacts with a range of groups. This finding suggests the relevance of informal, reflexive mechanisms to the ways in which state agencies implement policy "in the shadow of positive law." And the finding can assist in the evaluation of policy initiatives, such as EPA's recent emphasis on public-private partnerships. Thus, attention to the informal, reflexive aspects of state regulatory processes may prove useful to evaluators and other policy scientists concerned with how, and how well, states address their increasing responsibilities in the 1990s.

References

Babbie, E. *The Practice of Social Research*. (4th ed.) Belmont, Calif.: Wadsworth, 1989.

Carroll, J. D., Flynn, R. J., and Dorsey, T. A. "Vertical Coalitions for Technology Transfer: Toward an Understanding of Intergovernmental Technology." *Publius*, 1979, *9*, 3-33.

Feller, I., and Menzel, D. C. "Diffusion Milieus as a Focus of Research on Innovation in the Public Sector." *Policy Sciences*, 1977, *8*, 49-68.

Goldman, H., and Mokuvos, S. *The Privatization Book*. New York: Arthur Young, 1984.

Heilman, J. G., and Johnson, G. W. *The Politics and Economics of Privatization*. Tuscaloosa: University of Alabama Press, in press.

Ingraham, P. W. "Toward More Systematic Coordination of Policy Design." *Policy Studies Journal*, 1987, *15* (4), 611-628.

Johnson, G. W., and Heilman, J. G. "Diffusion of Innovation: Privatization of Municipal Wastewater Treatment." Paper presented at the annual meetings of the American Political Science Association, Chicago, September 3-6, 1987a.

Johnson, G. W., and Heilman, J. G. "Metapolicy Transition and Policy Implementation: New Federalism and Privatization." *Public Administration Review*, 1987b, *47* (6), 468-478.

McGarity, T. O. "Regulatory Reform and the Positive State: A Historical Overview." *Administrative Law Review*, 1986, *38* (4), 399-425.

Meier, K. J. *Regulation: Politics, Bureaucracy, and Economics*. New York: St. Martin's, 1985.

Miller, D. C. *Handbook of Research Design and Social Measurement*. (2nd ed.) New York: David McKay, 1970.

Rogers, E. M. *Diffusion of Innovation*. New York: Free Press, 1983.

Savage, R. L. "Diffusion Research Traditions and the Spread of Policy Innovations in a Federal System." *Publius*, 1985, *15* (4), 1-27.

Thompson, F. J., and Scicchitano, M. J. "State Enforcement of Federal Regulatory Policy: The Lessons from OSHA." *Policy Studies Journal*, 1985, *13* (3), 591-598.

Treiber, H. "Crisis in Regulatory Reform: Remarks on a Topical Theme; or, Reflexive Rationality in the Shadow of Positive Law." *Contemporary Crises*, 1985, *9* (3), 255-280.

U.S. General Accounting Office (GAO). *Water Quality: An Evaluation Method for the Construction Grants Program*. Report to the administrator of the Environmental Protection Agency. Document No. GAO/PEMD-87-4A&B. Washington, D.C.: GAO, 1987.

Weiss, C. H. "Evaluation for Decisions: Is Anybody There? Does Anybody Care?" *Evaluation Practice*, 1988, *9* (1), 5-19.

Welch, S., and Thompson, K. "The Impact of Federal Incentives on State Policy Implementation." *American Journal of Political Science*, 1980, *24* (4), 715-729.

Wise, C. R. "Public Service Configurations: Public Organization Design in the Post-Privatization Era." *Public Administration Review*, 1990, *50* (2), 141-155.

John G. Heilman is associate professor of political science at Auburn University, Auburn, Alabama. He has served since 1984 as a member of the organizing committee for the National Energy Program Evaluation Conference. Together with Gerald W. Johnson, he has written a forthcoming book on the politics and economics of privatization and is completing a book on state revolving loan funds.

Gerald W. Johnson is associate professor of political science at Auburn University, where he has served as head of the political science department, associate dean for research in the college of liberal arts, and chair of the university senate and general faculty. He has also served as an adviser to the governor, the legislature, and the supreme court of the state of Alabama.

The policy process that takes place in the siting of a high-level nuclear waste repository is marked by political controversy, an open decision-making process, scientific uncertainty, conflicting expert opinion, and high risk in case of system failure. Mandated quality-assurance studies of proposed sites represent high-stakes, high-pressure evaluation that must take a variety of factors into account.

Conflicting Expertise and Uncertainty: Quality Assurance in High-Level Radioactive Waste Management

Michael R. Fitzgerald, Amy Snyder McCabe

Management of the nation's high-level radioactive waste—both military waste from defense facilities and civilian waste from nuclear power plants—has traditionally been the purview of the federal government. Little attention was paid to disposing of possibly harmful residuals of nuclear energy production until the late 1950s, and since then the goal of permanent waste disposal has eluded federal policymakers. Lack of consensus as to the optimal method of disposal, disagreements over whether to combine military with commercial waste, and state and public outcries over exclusion from previous decisions that determined how and where to site nuclear waste facilities all led to a policy stalemate by 1980. Legislation enacted in late 1982, however, significantly altered the course of national waste-management policy, as Congress established a framework for siting, constructing, and licensing geological repositories to dispose of accumulating waste. After five years of mounting frustration and a slipping schedule, congressional amendment of the legislation in late 1987 designated Yucca Mountain, Nevada, as the potential host for the nation's first permanent geological repository for high-level nuclear waste.

In a review of American atomic energy regulation, Zimmerman (1988, p. 64) states that "relative to high-level radioactive wastes, the [disposal] problem apparently will be solved if tests demonstrate that the Nevada site will be a safe one." A major step for the Department of Energy (DOE) in making this determination is to implement a work plan, called a site characterization plan (SCP), that details the five-to-seven-year surface and underground examination of Yucca Mountain. Because information gathered during site characterization will be used to support construction and operating licenses for the repository, the Nuclear Regulatory Commission (NRC) requires that DOE develop and implement a quality-assurance (QA) program. The QA program, still under development at the writing of this chapter, is intended to ensure that data from the Yucca Mountain investigation are collected under strict quality controls. In this chapter we explore the dynamics of this large, expensive, and controversial evaluation effort by examining the factors that not only make QA a necessary undertaking but also complicate profoundly the utilization of its results. In part, the story told here illustrates the dilemma of high-stakes evaluation.

Quality Assurance in High-Level
Radioactive Waste Management

Although many factors affect siting decisions, one of the most basic in terms of acceptability is the degree of stakeholder confidence that a nuclear waste storage facility can be constructed and operated safely. This factor is especially salient in the siting of high-level nuclear waste facilities, as demonstrated by federal attempts to site a pilot repository in 1972 and a monitored retrievable storage (MRS) facility in 1985. These two examples underscore the point.

The first example involves DOE's predecessor, the Atomic Energy Commission (AEC). It was unable to defend the technical adequacy of a proposed site near Lyons, Kansas. Because the agency lacked scientific evidence for its claims that the facility would safely isolate radioactive waste, state experts were able to present their own evidence, forcing the AEC to cancel its project.

In the second example, a 1985 campaign to site an MRS facility in Tennessee was aborted for other than safety-related reasons. The Clinch River Task Force in Oak Ridge, Tennessee, a committee of local citizens, concluded that a MRS facility could be constructed and operated safely in their community, based on a systematic review of DOE's technical reports as well as the task force's own technical assessment. The state of Tennessee, like the city of Oak Ridge, was quick to organize its own review team to gather scientific evidence that would validate or repudiate DOE's data. At the end of a nine-month period during which the state evaluated the MRS, Tennessee's governor vetoed the DOE proposal, but not because the facility would have posed a threat to public safety. The governor said that despite the potential for safe MRS operation, he felt that the proposed facility was an unnecessary and expensive component of a national high-level waste-management system, and that its construction would stunt economic growth in the Oak Ridge area. Thus, even though the proposal for a MRS in Tennessee was eventually scrapped, the conclusions reached by the state and the local task force, based on their respective assessments of the available scientific evidence, in fact bolstered DOE's argument that a MRS facility could be a safe depot for spent fuel and high-level radioactive waste (see Fitzgerald and McCabe, 1988; McCabe, 1987).

These two examples suggest that valid and reliable scientific evidence is an essential precondition for establishing facility safety and, ultimately, stakeholder acceptance of a siting decision. Rigorous QA is thus far more than a formal element of the repository-licensing process; it is the central evaluative mechanism for demonstrating to state and local governments, the public, and to other stakeholders that the technical evaluation of the proposed site is based on accurate scientific information. This standard is shared by NRC, as reflected in a statement made by one NRC commissioner to Ed Kay, former acting director of the DOE Office of Civilian Radioactive Waste Management (OCRWM), in their first face-to-face meeting: "QA is occasionally neglected—at least in this agency. It's not enough in this business to do it right. You've got to be able to show it, to prove to the public that it was done right."

The task of establishing confidence in the technical quality of the proposed Yucca Mountain repository is no small undertaking for the federal government. Nevada officials have strongly opposed the siting at Yucca Mountain from the outset, claiming that political expediency, not technical superiority, was the primary consideration in the selection process. The state shows every indication of continuing its opposition during the licensing phase of the project, a process that will take at least until the mid-1990s to complete. A recent U.S. General Accounting Office (U.S. GAO, 1988, p. 11) report supports this observation: "Because NRC anticipates that the repository's licensing application will be heavily contested by the state of Nevada, among others, it is important that DOE's quality assurance program be designed and implemented in accordance with NRC's regulations. In NRC's licensing experience, uncontested and contested proceedings show a marked difference in the degree to which applicants have had to defend the quality of their work. In contested proceedings for nuclear power plants, if the applicant had quality-related weaknesses, interveners were successful in surfacing the problems during licensing. As a result, plant projects were canceled or incurred expensive and time-consuming delays while weaknesses were corrected."

During the licensing phase, Nevada could effectively stall or aid in the disqualification of Yucca Mountain by challenging scientific evidence presented by DOE and its contractors, or by claiming that the agency failed to follow approved QA procedures. It would then be up to DOE to prove that its QA program was properly executed and that public health and safety in the construction and operation of the waste facility would not be compromised.

For students of science-and-technology policy, DOE's current situation invites the question of whether a high-level nuclear waste repository can be rendered both scientifically sound and politically acceptable within the statutory time frame. For students of evaluation, the companion question has to do with the role, nature, and limits of QA. That is, what does this case show us about evaluation in a policy setting marked by high risk, complex technology, political controversy, and conflicting expert opinion?

Since the licensing process will continue until at least 1994, these questions cannot yet be answered with certainty. It is already possible, however, to develop foundations for the answer that will be sought. The present chapter is an attempt to do that by mapping out some of the factors that will strongly shape QA and the role that its results will play. Specifically, we systematically examine DOE's high-level radioactive waste program as an exceptional case of environmental policy, in which QA plays a vital role in program implementation. Our central theme is that for DOE a well-structured and administered QA program is absolutely essential to encourage stakeholder acceptance of scientific evidence, a key step to successful siting and licensing of a nuclear waste repository in Nevada. We contend that while QA programs are essential to the development and implementation of major environmental policies in the presence of scientific uncertainty, conflicting technical expertise, and political controversy, such programs cannot guarantee that policy goals will be attained. This situation is of concern because an unsuccessful QA program will deliver a severe, if not fatal, blow to the implementation of the nation's first comprehensive radioactive waste management policy.

In the following sections of this chapter, we introduce QA as an evaluative process and then examine DOE's QA program to date. We then review some of the major factors that simultaneously contribute to the need for, yet potentially threaten the effective utilization of, a first-rate DOE QA program.

What Is QA? Although the QA program for the Nevada repository site is still under development, much can be said about the process in general. The concept of QA is familiar to many scientists, yet its application varies across disciplines. In the engineering sciences, for example, QA regulations are quite commonly applied to facility construction where engineers can manipulate their environment to a considerable extent. As an operating principle, engineers document their work for traceability purposes; provision of such documentation is often necessary in order to obtain construction and operation licenses from government entities. Other scientists take a more orthodox approach to QA, as expressed by the following National Research Council Colloquium "laboratory" definition:

> Quality assurance from the laboratory perspective consists of an integrated system of procedures. The procedures interface in such a way that the progress of the work is tracked and documented so that at the

end of the work, one can ascertain from the documentation what was accomplished, and how successfully it was accomplished. There are four primary elements to the laboratory quality assurance system: activity plans, accurate measurement, traceability, and storage [Senseny, 1989, p. 4].

Among the major advantages, then, of laboratory QA procedures are that they enhance reproducibility and confidence in the data. In this regard, they serve to "confirm" the quality of data and analysis somewhat in the manner suggested by Lincoln and Guba (1985).

Private- and public-sector organizations have employed QA as a management tool for many years. Manufacturers use a variety of QA methods and procedures to monitor uniformity in the quality and durability of their products. Historically, this approach to QA has been limited to inspection and rejection of manufactured goods (Claudson, 1989, p. 1). But the Japanese concept of total quality management has prompted a change in traditional Western managerial styles. Japan's experience with QA maintains a total organizational commitment to "quality first," not "profit first." In other words, not only are inspectors responsible for quality control, but also product designers and manufacturers and, ultimately, top management assume responsibility for customer satisfaction (Ishikawa, 1985, p. 76).

In the public sector, federal agencies such as the National Aeronautics and Space Administration and the Department of Defense require that companies from whom they procure products comply with governmental QA regulations. In contractual arrangements of this sort, the purchasing agency is interested in its suppliers' ability to consistently produce quality goods (American National Standards Institute/American Society for Quality Control, 1987, p. 2). In the area of environmental policy administration, QA is critical to the implementation of major waste-management policies. Agencies such as the Environmental Protection Agency (EPA) and the DOE, whose missions encompass hazardous, toxic, and radioactive waste remediation, must be confident that the data defining the source and extent of contamination are highly dependable, as well as defendable in court proceedings. The EPA's Superfund program, which was created by litigation, relies on accurate risk assessments to help determine existing public health risks, and also to establish cost and liability for environmental restoration (Hammond, 1989, p. 1).

To this effect, the EPA developed Quality-Assurance Program Plans as part of an effort to improve its data collection and evaluation processes. The EPA quality-improvement process stresses the participation of all employees, especially top management, as well as training, reliance upon standards and measures of progress, and rewards and recognition as keys to success (Collins, 1988, p. 43). Similarly, the DOE's newly revised order 5820.2A, governing radioactive waste management, contains prescriptive requirements for the gamut of radioactive materials handled by the agency and reiterates that waste-management operations must adhere to established QA standards. DOE, in its SCP of December 1988, states that QA consists of "all the planned and systematic actions necessary to provide adequate confidence that a structure, system, or component is connected to plans and specifications and will perform satisfactorily" (U.S. DOE, 1988b, p. 144). Implementation of these broad guidelines has not been without problems, as can be seen in a review of DOE's experience in practice with its contractors and with other federal agencies.

DOE QA Activities to Date. DOE has conducted experiments at Yucca Mountain since 1977, prior to the enactment of the Nuclear Waste Policy Act and long before a formal QA program was mandated, let alone developed and approved. Because much of the research was not performed under strict quality controls, DOE made the determination in 1986 to order some of its contractors to cease work at Yucca Mountain until proper QA programs were implemented. DOE's decision was based on its assessment that control over work on the Yucca Mountain project was inadequate for NRC licensing purposes, and that since the Nevada site was selected that year as a finalist for the proposed repository, improvement was not proceeding at a satisfactory pace in preparation for full-scale site characterization (U.S. GAO, 1986, appendix 1). Thus, between April 1986 and December 1987, six contractors to the Nevada DOE office were issued stop-work orders until corrective measures were taken regarding audit findings, and until the project office approved QA program plans.

There followed a barrage of criticism aimed at DOE concerning the lack of a well-defined and institutionalized QA program for the Yucca Mountain investigation. The NRC in its review of the draft SCP found— and the U.S. GAO (1988, p. 3) concurred in a report to the chairman of the House Subcommittee on Energy and Power—that there was "insufficient basis at this time for confidence in the adequacy of DOE's quality assurance program." And in an unusual public display of skepticism, the Edison Electric Institute/Utility Nuclear Waste Management Group (EEI/UNWMG), in a letter to a top DOE manager, wrote that "EEI/UNWMG have been concerned for a long time over the lack of progress in developing Quality Assurance plans and procedures for the DOE repository program."

Responding to a NRC comment that QA regulatory standards were not being met, the DOE established a separate QA office in April 1988. Further organizational changes were ushered in with the newly installed QA management, the most important of which is that the QA managers currently report directly to the OCRWM director. This action resolved a NRC concern that the DOE senior QA management position was too low in the organizational structure to be effective. Another modification is that QA managers and staff are now located at headquarters, at the Yucca Mountain project office, as well as at project contractors' offices. Furthermore, in its 1988

Draft Mission Plan Amendment (U.S. DOE, 1988a, p. 71), DOE discusses its 1987 redefinition of QA policies in line with the OCRWM director's "managing for quality" program. Under this mandate, all activities pertaining to radiological safety and selected non-safety-related activities are subjected to QA provisions established for nuclear facilities by the American National Standards Institute and the American Society of Mechanical Engineers. This policy defines a theme discussed later here: As controversy mounts over programs and their evaluation, the policy process tends to be reorganized to include increasing numbers of stakeholders from both the public and the private sectors. The complexity of the situation is evident in the mix of factors that both make QA indispensable and drastically complicate its effective use.

Need for an Effective QA Program

In addition to the regulatory requirements with which DOE must comply, a number of other factors make QA an indispensable—and problematic—component of the nuclear waste siting process. We examine six of these factors here and briefly discuss how each relates QA to the Yucca Mountain program.

Political Controversy. That politics inevitably plays a substantial role in scientific and technical policy-making is a major theme throughout a robust literature on the subject (see, for instance, Kraft and Vig, 1988; Goggin, 1986; Dickson, 1988; Barke, 1986). From the earliest stages of agenda setting to the later stages of resource allocation, political institutions and processes dominate the formation, implementation, and evaluation of science-and-technology policy. There is considerable debate, however, over how large a role politics does play, and ought to play, in decision making where, as in the case of nuclear waste management, there exist not only technical concerns but ethical, philosophical, and equity issues as well.

The search for a solution to the nation's high-level radioactive waste predicament has been almost thirty years in the making (comprehensive accounts appear in Colglazier and Langum, 1988; Carter, 1987). Along the way, the process has been riddled with accusations that siting decisions were made based on political rather than technical criteria. The most recent attempt to site a repository in Nevada is no exception and, in fact, may yet prove to be the most intense battle over state's rights since the Civil War.

A review of events of the past decade tells the story. Recognizing the long-overdue need for a comprehensive, national, radioactive waste management policy, Congress, after many aborted attempts, passed the Nuclear Waste Policy Act in late 1982. In 1985, DOE selected three sites in Tennessee as potential, temporary waste storage facilities, and in May 1986 the agency added Texas, Nevada, and Washington as finalists for the first geological repository. The selection criteria used by DOE were challenged in court by the finalist states, which claimed that the choices were made for political, not scientific, reasons. Consequently, strained relations between the federal, state, and local officials thwarted DOE's siting efforts. These factors, along with an opposed citizenry in the affected areas, led to congressional amendment of the act in December 1987. The Nuclear Waste Policy Amendments Act (NWPAA) designated Yucca Mountain, Nevada, as the only site that would be scientifically evaluated as a potential repository, thereby abandoning the original three-site strategy. This congressional action in effect declared that the comparative site selection process set forth in the original legislation was a bankrupt effort.

Contentiousness, rather than cooperation, has been the hallmark of federal-state relations since the passage of NWPAA. Nevada officials contend that congressional leaders from Washington, Texas, and Louisiana, in a typical quid pro quo fashion, "ganged up" on their state in December 1987, replacing a scientific process with a purely political one. And members of Congress did little to hide the political maneuvering that was behind the decision, nor did they show any meaningful sign of remorse. As one Democrat from Washington declared, "We've done it [selected Nevada] in a purely political process. . . . We are going to give somebody some nasty stuff" (Davis, 1987). Senator Bennett Johnston of Louisiana, the man who engineered the House-Senate negotiation, added, "If I were a Nevadan living in the real world, I would be happy with this bill. I would bet that in very few years, Nevada will deem this one of their most treasured industries." When Nevada legislators complained that they were locked out of the conference that produced the agreement, Johnston responded, "They [Nevadans] weren't shut out . . . they just weren't appointed to the conference" (Davis, 1987).

It is small wonder that Nevadans have resorted to similar tactics. Their most recent attempts to stave off the repository include the passage of a bill by the Nevada General Assembly on June 28, 1989, making it unlawful for the federal government to store nuclear wastes at Yucca Mountain, as well as substantially delaying the state's issuance of air quality permits needed by DOE to begin drilling at Yucca Mountain. For its efforts, the state received the news in July 1989 that a Senate appropriations subcommittee, also chaired by Johnston, authorized the withholding of $6 million in funding for Nevada's oversight program for fiscal year 1990. The subcommittee directed DOE to withhold the money until DOE's Secretary Watkins certifies "good faith efforts and cooperation by the state to permit federal tests for the nuclear waste facility." Johnston cited the new Nevada law banning repositories as the other reason for the new restriction (Adams, 1989a, 1989b).

Although politics has entered into what was originally acclaimed to be a scientific enterprise, DOE maintains that a disposal facility will not be constructed at Yucca Mountain unless scientific evidence gathered during site characterization supports preliminary assessments as to the integrity of the site. For DOE, in light of congressional behavior of late, nothing less than a perfect QA program can come close to appeasing the state at this stage in the siting process. That is, if actions in Congress since passage of the amendment are any indication of how decisions will be made in the future, DOE has an enormous, perhaps insurmountable, challenge ahead. For the state there is cause for concern as well; it is the U.S. Congress that will determine the final selection or rejection of Yucca Mountain.

More Open Decision Making. During the 1970s, congressional action and a series of judicial decisions modified the processes by which environmental policy is made in the United States in terms of who should participate and how much they should know (Jasanoff, 1986, pp. 71–76). Legislation and agency rule making provided avenues for the public to play a larger role in environmental decision making. State and local governments, at least partly due to the Reagan administration's efforts to reshape and reform American federalism, have also become more active in the policy process by assuming many environmental decision-making and implementation responsibilities that were traditionally the exclusive domain of their federal counterpart (Fitzgerald, McCabe, and Folz, 1988, p. 98).

This trend toward more open environmental decision making has important ramifications for policy implementation and program evaluation. On the one hand, a better-informed, more involved public (and subnational governments) may lead to better decisions and can be viewed as increasing the legitimacy of policy choices made in a democratic society. On the other hand, facilitation of access to decision making can be viewed as an administrative and regulatory nightmare. That is, the larger the number of participants trying to resolve a particular conflict, the harder the task becomes for those in charge. Schattschneider (1960, p. 3) ably describes the result: "So great is the change in the nature of any conflict likely to be as a consequence of the widening involvement of people in it that the original participants are apt to lose control of the conflict altogether." One could argue that a widened scope of conflict in nuclear waste siting greatly increases the need for a QA program of the highest caliber.

In contrast to the implementation of nuclear programs of previous decades, the Nuclear Waste Policy Act of 1982 mandates, and provides funds for, a review by subnational governments and Indian tribes of federal agency performance. The act also mandates public review of agency documents and requires DOE to hold a series of public hearings at various stages in the siting process. Contracts negotiated in 1983 between nuclear utilities and DOE established a nuclear waste fund to pay for the repository program, in exchange for the federal government's agreement to accept title to the waste by 1998. The state of Nevada receives appropriations from the nuclear waste fund, authorized by Congress, to oversee DOE management of the nuclear waste project at Yucca Mountain. Nevada's

Agency for Nuclear Projects, exercising the oversight function, has grown rapidly over the past several years, employs its own in-house technical staff, issues a newsletter periodically, and supports a small public information agency with money supplied by DOE.

Schattschneider's prophecy has been realized in Nevada; as the number of participants in the nuclear waste siting controversy has increased, the scope of the conflict has expanded. And for DOE to administer a program under these circumstances is a task envied by no other federal agency. The state, for example, has used DOE grant funds to fight the repository and has taken advantage of every speaking engagement and hearing to publicly voice Nevada's opposition to the repository. In fact, in response to a request by key congressional members, GAO undertook an audit to ensure that Nevada has dispensed funds for authorized purposes only, rather than for payment of Washington-based lobbyists allegedly retained by the state to fight the siting decision. A number of other information groups, some supported by the nuclear industry and some staffed by environmentalists, supply information pertaining to Yucca Mountain. There are, therefore, reams of information available to the public, and a number of opportunities available for those who wish to become actively involved in the siting controversy.

Because of more open decision-making system in nuclear waste siting policy, that is, a widening in the scope of conflict, an exceptional QA program at Yucca Mountain is critical to program implementation. The oversight of DOE's repository project by Congress, NRC, state and local governments, and the public increases the need for an effective QA program, since these groups have opportunities to challenge DOE's scientific claims concerning Yucca Mountain. In fact, the state and its allies continually label data supplied by DOE as suspect, adding weight to the argument for rigorous QA standards. Thus, it is important to study and evaluate not only the impact of more open decision making but also the role of QA as it relates to the resolution of environmental disputes.

Organizational Effectiveness. The siting, construction, and operation of a permanent geological repository that must function effectively for thousands of years is surely among the most challenging tasks ever delegated by Congress to an executive agency. From the inception of the repository program, serious questions were raised concerning the organizational capacity of DOE in general, and its OCRWM in particular, to accomplish its mission. At virtually every stage of the process to date, OCRWM's organizational effectiveness and managerial competence have been challenged. Such criticism, in part, can be taken as a natural extension of DOE and predecessor agency failures in handling radioactive waste at defense and research facilities; that is, how can the organization that helped create the problem in the first place be trusted to implement a permanent solution?

A large amount of criticism, however, is leveled by a variety of agencies and groups specifically concerned with oversight of the deep repository project, all of which points to abiding problems in getting the program on track and keeping it there. A major impetus for the 1987 amendments of the Nuclear Waste Policy Act was, for example, congressional dissatisfaction with OCRWM's putative organizational inability to simultaneously proceed with site characterizations in three regions. Headquarters and project office reorganizations, turnover in staff and line personnel, shifting budgetary priorities, congressional redirection with the 1987 amendments, all have contributed to monumental problems in making OCRWM an effective management organization. Adding to existing problems is the fact that no permanent OCRWM director has been named since Ben Rusche left DOE in 1987.

It will be some time before observers can tell whether or not organizational changes initiated by the Bush administration will have an impact at the level of the Yucca Mountain project. The new secretary of energy, Admiral James Watkins, appears to be employing a more centralized management style at DOE, as indicated by his creation of the office of assistant secretary for waste management, under which all civilian and defense waste programs fall, and whose appointee reports directly to the secretary. Moreover, under Watkins DOE is placing great emphasis on QA as an instrument for creating and demonstrating organizational effectiveness.

Now that the Yucca Mountain project is the focus of the repository program, concern centers on whether OCRWM can possibly develop an adequate QA program for such a difficult and controversial undertaking. In organizational terms alone, the task is indeed sobering to contemplate. The Yucca Mountain project, which is organizationally located within DOE's Nevada operations office rather than directly under OCRWM, in 1989 employed fourteen hundred scientists, engineers, and support personnel with a budget of over $200 million. But less than one hundred of the project staff are DOE employees; the rest are employed by eight separate contractors and various subcontractors, including personnel at National Laboratories and other federal agencies. To coordinate and oversee the efforts of this array of project elements, DOE, after substantial delay, finally named Bechtel as its "gorilla contractor" for the Yucca Mountain project; but the selection is being challenged in federal court. Citing inadequate QA program implementation, the Yucca Mountain project manager has issued its most recent stop-work order to the Sandia National Laboratory. The Yucca Mountain project in a sense has become almost entirely an effort to create a credible and effective QA program. With respect to the Nuclear Waste Policy Act then, it appears as if DOE's organizational effectiveness in general, and that of OCRWM in particular, rides largely on the QA program.

Scientific Uncertainty. In studying the role of expertise in technological societies, Benveniste (1972, p. 194) observed that "the more uncertainty there is, the more we hear a call for centralized and comprehensive planning." Clearly, the desire to reduce uncertainty, and thereby reduce—it is hoped—risk in the disposal of the by-products of nuclear power, is a powerful force behind QA programs. Indeed, perhaps the most nagging problem associated with the deep repository program is the scientific uncertainty that permeates this congressionally mandated approach to disposing of spent nuclear fuel. For opponents, most especially the state of Nevada, the question is not simply *do* we know enough, but *can* we know enough—given the present state of the relevant scientific and technical disciplines—to safely store this material for thousands of years? Are we capable, at least for the foreseeable future, of satisfactorily knowing whether the proposed permanent solution to storing high-level nuclear waste can really work?

Much of the project-related research occurring around the country is devoted to assuring that all of the elements of the program, such as packaging, containment, and transportation, will function properly. The site characterization process itself is specifically predicated on the assumption that we presently do not know nearly enough about Yucca Mountain to be scientifically "certain" (that is, able to predict what will happen on the site within acceptable limits of probability and risk) that nuclear waste can be permanently stored there safely for thousands of years. Dickson (1988) has observed that, as part of the "new politics of science," science can be used to legitimize desired policy goals. Obviously, a successful QA program can further bolster the weight of science as legitimation. Whether the DOE program actually can legitimize the decision to opt for deep geological storage as a permanent solution to the high-level nuclear waste problem is an open issue. And the outcome will be far-reaching in terms of evaluation: If the science involved disqualifies Yucca Mountain, the DOE program will likely fatally subvert an already beleaguered approach.

Conflicting Expertise. When it comes to public acceptance of a high-level nuclear waste repository, who says how safe is safe enough, and can QA assure it? In other words, whose word is sufficient to confirm that QA procedures ensure that the data collected are valid and reliable? When policy choices must be made in the presence of scientific uncertainty, those claiming expertise rarely agree on either the appropriate research methodologies or the level of risk associated with a particular technology. In a discussion of risk assessments of hazardous substances, Jasanoff (1986, p. v) observes that experts frequently bias their investigations in favor of their own belief systems: "In the absence of definite knowledge, expert opinions tend to be colored by personal values as well as professional judgment, leading to different assessments of the significance of particular risks. In the effort to manage risks, public authorities are thus drawn into mediating not only among competing economic and political interests, but also among conflicting technical interpretations informed by widely divergent views about pollution, nature, illness, and death."

Regens and Dietz (1985, pp. 6-7) add, moreover, that technical expertise lends itself to a discretionary power that enables experts to help set the boundaries for a policy debate; expertise, then, becomes a source of political power. Because public administrators are often arbiters among competing interests, and because expertise is a source of political power in technical decision making, it is essential that QA is practiced by those whose research is employed for decision-making purposes. For studies conducted of potential nuclear waste sites, QA programs thus become extremely valuable. The brief history of the Yucca Mountain project has produced two major incidents where conflicts among technical and scientific experts have occurred, and DOE will have to respond to both. The first deals with a contractor, the other with separate reports issued by DOE and NRC scientists.

Some geologists employed by the U.S. Geological Survey (USGS), a major contractor to DOE at the Nevada site, are concerned that DOE's current QA program does not appropriately address the needs of earth scientists. This group within the agency contends that while QA practices are appealing in the laboratory setting, the successful application of QA regulations to field experiments is somewhat limited. Some of the USGS scientists argue that the DOE-imposed QA requirements do not fit well within the structure of research activities of geologists and hydrologists. It can be extremely difficult for these scientists to manipulate their environments, and the sheer cost and amount of time required to plan and document procedures renders QA impractical for certain types of in situ testing. The experimental underground work necessary for determining the suitability of Yucca Mountain requires sophisticated, customized geotechnical instrumentation.

Any program imposed on instrumentation, therefore, must consider that alterations will likely be needed during the experimental phase. Therein lies one of the controversies with the application of QA to repository site characterization: How does DOE design and implement a QA program imposing a priori acceptance criteria for experimental processes in which unanticipated changes in geological conditions must be accommodated? In other words, could QA as applied by government agencies accommodate the approach of experimental scientists engaged in an unprecedented undertaking?

A major complaint of the USGS scientists is that the restrictions imposed by DOE's QA procedures will discourage creative research and cause problems in recruiting talented investigators. These concerns culminated in the submission of an August 17, 1988, memorandum to an assistant chief hydrologist at USGS by sixteen hydrologists and hydrologic technicians employed by the Yucca Mountain project office. The authors complained that "it is also generally recognized that our current quality assurance (QA) program is modeled after the nuclear power industry's reactor facilities QA guidelines. As a result, the present QA program is engineering oriented, inappropriate in most instances, and counterproductive to the needs of good scientific investigations. There is no facility for trial-and-error, for genuine research, for innovation, or for creativity."

The problems for DOE, in terms of conflicting expertise, do not end with a resolution of the USGS dispute. The issuance of two reports challenging the integrity of the Yucca Mountain site place the agency in a sensitive, some would say defensive, position. In 1987, then-governor Richard Bryan of Nevada released a report authored by a DOE geologist that claims that volcanic and earthquake activity near the proposed site has not been adequately examined. The scientist's report suggests that a shift in the tuff at Yucca Mountain could cause the water table to rise and infiltrate the repository constructed within; he concludes that "my personal view is that Yucca Mountain, having these conditions that must be substantiated and verified, is essentially unlicensable" (Shetterly, 1989).

Similarly, a senior geologist at NRC in a memo to his superiors concluded that "I personally believe that the Yucca Mountain site should be dropped from consideration for a nuclear waste repository" (Maize, 1989). The NRC scientist based his conclusions on data gathered by the University of Nevada, Las Vegas, that projects a possible increase in volcanic activity over the course of the minimal containment period of ten thousand years. These studies repudiate previous conclusions drawn by DOE geologists and have been cited by repository opponents as further evidence that nuclear waste disposal at Yucca Mountain could prove to be a dangerous enterprise. Responding to the conflicting studies, DOE has agreed to accept an independent review of the DOE scientist's report, and to investigate the concerns cited by the NRC geologist. While DOE and NRC maintain that these studies do not represent official agency positions, each acknowledges that the issues must be resolved.

The authors of these two evaluations, in effect, have helped set the boundaries for debating the siting process at Yucca Mountain. QA will play a major role in DOE's response, as the agency seeks to verify the data on which the reports are based. No matter whether the agency replicates the studies or presents contradictory evidence, DOE will have to justify its conclusions with scientific evidence. Likewise, DOE will have to determine what constitutes QA for earth science investigations. If DOE imposes a QA program on the USGS that does not, in the opinion of the geologists, accommodate the needs of creative research, the program could incur a substantial delay while USGS revises and implements new QA plans and procedures. Conflicting expertise, therefore, adds to both the need for and the complexity of rigorous QA in nuclear waste siting. DOE must ultimately convince NRC and other stakeholders that its conclusions are based on scientific evidence backed by QA, and that the evidence can stand up against competing claims made by other scientists.

Consequences of System Failure. The need and demand for QA programs has dramatically increased in recent years because the consequences of system failure are more obvious and severe. Experience has proved a hard teacher in this regard, especially with respect to applied science, technology, and the environment. Our inability to anticipate the long-term health and environmental consequences of a failure to control the manufacture, use, and disposal of chemicals and radioactive materials, for example, has yielded social, political, and environmental problems of escalating proportions. The keystone of the elaborate regulatory structure that is evolving in response to these problems is quality control: creating, building, maintaining, and upgrading systems properly so as to prevent system breakdown.

Fueled by a mass communications process that exacerbates the growing sense that humans and the ecosystem can, and do, suffer greatly when control technologies break down, the imperative to ensure that controls really will work, before they are fully operational on a large-scale basis, increasingly defines the regulatory environment. At base this is a concern for what happens if things are not designed, built, operated, and maintained properly, but the issue includes as well a fear that the scientific and technical knowledge on which systems engineering is based is insufficient to ensure that there is effective control for safe operation—as well as for effective amelioration when things go wrong.

Summary and Conclusion

The problems associated with radioactive waste siting policy in the United States are symptomatic of a larger dilemma that contemporary environmental policymakers regularly encounter: Who, in a democratic society, should pay the costs, bear the risks, and enjoy the benefits associated with a particular technology, and how should the choice be made?

Congress passed the Nuclear Waste Policy Act to set up a process by which the nation's first high-level nuclear waste repository could be sited, licensed, constructed, and operated. An amendment to the act in late 1987 designated Yucca Mountain in Nevada as the only site where DOE is authorized to perform the scientific and technical assessments necessary to determine the site's suitability for radioactive waste isolation. Under the provisions of the Nuclear Waste Policy Act, NRC requires DOE to subject all scientific studies conducted by the agency and its contractors to rigorous QA standards.

In addition to the legal parameters that bind DOE to the implementation of a NRC-approved QA program, other factors contribute to the need for employing QA practices in nuclear waste facility siting. This study sought to examine each of these factors—scientific uncertainty, a more open decision-making system, political controversy, organizational effectiveness, conflicting expertise, and potential consequences of system fail-ure—in order to establish the role of QA in high-level radioactive waste siting policy. Based on this review, what conclusions can we draw?

First, given the myriad forces that increasingly define the environ-mental policy domain, continued—even increased—emphasis on QA is inevitable, but it does not come without substantial cost or danger. The development, implementation, and monitoring of the extensive and inten-sive QA program underway at DOE adds millions of dollars to program costs and inevitably means delays. The idea is that the long-term savings attributable to a better, safer, and more acceptable program will offset such costs, or at least involve lower costs (in the long run) than those generated by a less rigorous (let alone nonexistent) program. Whether, in practice, this balance can or will be achieved remains to be seen and must be tracked.

Major organizational costs will inevitably be incurred by the burgeon-ing QA program. Stop-work orders engendered by QA requirements, how-ever necessary, are certain to affect morale and resource allocation in both the short and long terms. Key personnel and resources can be lost during periods of inactivity, especially if down times prove prolonged. That is, if in the pursuit of its QA program DOE is inattentive to the effect of the program on the ability of its line managers to recruit, and maintain in productive work environments, first-rate scientific and technical personnel, the program could result paradoxically in a weakened or less able imple-mentation system. An excessively strict QA system could engender frustra-tion, resentment, and alienation among key personnel who, over time, simply exit the organization to ply their crafts and trades in more satisfac-tory environments.

Second, QA in its broadest purpose is supposed to control the discre-tion of those within the system in order to ensure that their missions are accomplished properly. This control is intended to ensure that even years after thousands of specific tasks are completed, the work can be demon-strated to have been done correctly. Yet, the same critical concern that Byner (1987, p. 218) applies to efforts to control bureaucratic discretion in general, applies to scientific and technical discretion in particular: "The debate over control of the bureaucracy that has characterized the last decade should be balanced with a concern for bureaucratic competence. Bureaucratic discre-tion cannot be understood apart from the tasks that agencies are delegated to accomplish. Excessive and misdirected actions to reduce discretion reduce the capability of the administrative process to accomplish its dele-gated tasks." Thus, systematic attention needs to be paid to exactly how, and to what effect, the balance between control and competence, as manifested in the DOE QA program, is actually struck over time. In this regard, then, QA is largely but a scientific and technical manifestation of the unending con-cern for accountability in a democratic system.

Third, it is by no means certain that even a "perfect" QA program can ensure the level of acceptance that will be required by regulatory agencies, executive and legislative decision makers, key elites, and the public in the future. It is the nature of the future, especially if the past is prologue in any program dealing with radioactive material, to be uncertain. Thus, even if DOE is able to execute a first-rate QA plan that accommodates creative research, and can allay all other concerns about its QA program, what remains uncertain is the agency's ability to overcome the scientific uncertainty and political controversies that continue to envelop the entire repository project. The state of Nevada has consistently argued that Yucca Mountain does not possess the geological characteristics that would make the site suitable for isolating high-level radioactive wastes from the environment, and reports prepared by one DOE and one NRC geologist appear to support the state's contention.

Fourth, a potential "danger" inherent to QA programs, especially the one associated with nuclear waste disposal, then, is that they amount to futile attempts to accomplish goals that QA alone cannot possibly achieve: public acquiescence and political consensus for the program. For example, it is at least possible, and is perhaps even likely, that a rigorous QA program will provide the basis for challenging as well as supporting DOE's waste disposal mission. Further, in the absence of the discovery of a "fatal" flaw during site characterization, it will be the president and Congress that ultimately decide the fate of the Yucca Mountain project. As with all public policy decisions regarding matters nuclear, a variety of factors—most of which are at best tangential to the kinds of concerns addressed by QA programs—will affect presidential and congressional behavior. If in the design, implementation, and monitoring of its QA program DOE attempts too much, it may squander precious time and resources only to meet political defeat at the end of the long road. If it attempts too little, it will fail to meet the quality standards sufficient for regulatory approval prior to a final political sign-off.

Can a high-level nuclear waste repository be rendered both scientifically sound and politically acceptable? The answer at present is unclear. What is clear is that the science will definitely be challenged, and DOE must make its case not only before the U.S. Congress and NRC but also Nevada, the mass media, the public, and, ultimately, the courts. While even a perfect QA program may not "save" DOE's Yucca Mountain project, an unsatisfactory program will almost certainly doom it. How this beleaguered federal agency proceeds with QA in its deep repository program surely will tell us a great deal about the vicissitudes of implementing long-term environmental policies that are rooted in complex, unsettled, and controversial areas of science and technology. In particular, we can learn much about the organizational mixing of public and private sectors in this complex process, and about the conduct and role of QA as a form of evaluative research.

References

Adams, S. "DOE Denies Involvement in Fed Funding Proposal." *Las Vegas Review-Journal,* July 22, 1989a, p. 1A.

Adams, S. "State to Receive $4 Million for Yucca Mountain Studies." *Las Vegas Review-Journal,* July 27, 1989b, p. 1A.

American National Standards Institute/American Society for Quality Control (ASQC). *American National Standard: Quality Management and Quality Assurance Standards-Guidelines for Selection and Use.* Document No. Q90-1987. Milwaukee, Wis.: ASQC, 1987.

Barke, R. *Science, Technology, and Public Policy.* Washington, D.C.: Congressional Quarterly Press, 1986.

Benveniste, G. *The Politics of Expertise.* Berkeley, Calif.: Glendessary Press, 1972.

Byner, G. C. *Bureaucratic Discretion: Law and Policy in Federal Regulatory Agencies.* Elmsford, N.Y.: Pergamon Press, 1987.

Carter, L. J. *Nuclear Imperatives and Public Trust: Dealing with Radioactive Waste.* Washington, D.C.: Resources for the Future, 1987.

Claudson, T. T. "Leading the Spirit of Quality." Paper presented at the International Waste Management Conference of the American Society for Quality Control, Las Vegas, Nevada, April 2-5, 1989.

Colglazier, E. W., and Langum, R. B. "Policy Conflicts in the Process for Siting Nuclear Waste Repositories." *Annual Review of Energy,* 1988, *13,* 317-357.

Collins, F. C., Jr. "A New Era at the Environmental Protection Agency." *Quality Progress,* 1988, *21,* 43-44.

Davis, J. A. "Nevada to Get Nuclear Waste; Everyone Else 'Off the Hook.'" *Congressional Quarterly,* 1987, *45,* 3136.

Dickson, D. *The New Politics of Science.* Chicago: University of Chicago Press, 1988.

Fitzgerald, M. R., and McCabe, A. S. *The U.S. Department of Energy's Attempt to Site the Monitored Retrievable Storage Facility (MRS) in Tennessee, 1985-1987.* Knoxville, Tenn.: Energy, Environment, and Resources Center, 1988.

Fitzgerald, M. R., McCabe, A. S., and Folz, D. H. "Federalism and the Environment: The View from the States." *State and Local Government Review,* 1988, *20,* 98-104.

Goggin, M. L. (ed.) *Governing Science and Technology in a Democracy.* Knoxville: University of Tennessee Press, 1986.

Hammond, J. D. "Judicious Application of Quality Assurance Based on Value Added and Risk Encountered." Paper presented at the International Waste Management Conference of the American Society for Quality Control, Las Vegas, Nevada, April 2-5, 1989.

Ishikawa, K. *What Is Total Quality Control? The Japanese Way.* Englewood Cliffs, N.J.: Prentice-Hall, 1985.

Jasanoff, S. *Risk Management and Political Culture.* New York: Russell Sage Foundation, 1986.

Kraft, M. E., and Vig, N. J. (eds.) *Technology and Politics.* Durham, N.C.: Duke University Press, 1988.

Lincoln, Y., and Guba, E. *Naturalistic Inquiry.* Newbury Park, Calif.: Sage, 1985.

McCabe, A. S. "Open Systems of Environmental Decision Making: The MRS Nuclear Waste Siting Case in Tennessee." Unpublished doctoral dissertation, Department of Political Science, University of Tennessee, 1987.

Maize, K. "Volcanic Activity Makes Yucca Mountain Unsuitable as Waste Dump, Says NRC Geologist." *Energy Daily,* July 21, 1989, p. 1.

Regens, J. L., and Dietz, T. M. "Risk Assessment and Environmental Regulation: The Role of Technical Analysis in Policy Choice." Paper presented at the annual meetings of the American Political Science Association, New Orleans, Louisiana, August 29–September 1, 1985.

Schattschneider, E. E. *The Semi-Sovereign People*. New York: Holt, Rinehart & Winston, 1960.

Senseny, P. "Laboratory Testing." In *Quality Assurance Aspects of Geotechnical Practices for Underground Radioactive Waste Repositories: Proceedings of a Colloquium*. Washington, D.C.: National Academy Press, 1989.

Shetterly, C. "Yucca Mountain Project Foe Urges New Studies." *Las Vegas Review-Journal*, July 27, 1989, p. 4A.

U.S. Department of Energy (DOE). Office of Civilian Radioactive Waste Management. *Draft 1988 Mission Plan Amendment*. Document No. DOE/RW-0187. Washington, D.C.: DOE, 1988a.

U.S. Department of Energy. Office of Civilian Radioactive Waste Management. *Site Characterization Plan Overview*. Document No. DOE/RW-0198. Washington, D.C.: DOE, 1988b.

U.S. General Accounting Office (GAO). *Nuclear Waste: Repository Work Should Not Proceed Until Quality Assurance Is Adequate*. Washington, D.C.: GAO, 1986.

U.S. General Accounting Office. *Nuclear Waste: Repository Work Should Not Proceed Until Quality Assurance Is Adequate*. Document No. GAO/RCED-88-159. Washington, D.C.: GAO, 1988.

Zimmerman, J. F. "Regulating Atomic Energy in the American Federal System." *Publius*, 1988, *18*, 51–65.

Michael R. Fitzgerald is professor of political science at the University of Tennessee, Knoxville. He also serves as senior fellow at the University of Tennessee Energy, Environment, and Resources Center and as visiting senior fellow at the Center for Resource and Environmental Policy, Institute for Public Policy Studies, Vanderbilt University, Nashville, Tennessee.

Amy Snyder McCabe is research associate at the Energy, Environment, and Resources Center, University of Tennessee. She also serves as fellow at the Center for Resource and Environmental Policy, Institute for Public Policy Studies, Vanderbilt University.

AFTERWORD

The Editor's Notes and prior four chapters in this volume deal with privatization and its evaluation. The term *privatization* covers a broad range of strategies for service production. The unifying theme is that the role of the private sector increases relative to the role of the public sector. A growing literature in public administration and policy science addresses privatization. The terminology in this rapidly expanding area of research is in flux, and it is also tricky. Words make a difference here. The term *public-private partnership* is sometimes used instead of *privatization,* but both terms have been adopted by federally elected or appointed officials so that they are ideologically freighted. The terms *third-party government* and *government-by-proxy* are more neutral, and certainly useful, but they may imply some conclusions about what government is or should be.

The term *public-private configuration* serves several positive functions. First, the simple juxtaposition of the two labels, public and private, implies relationship, and the possibility of balance. Second, the notion of configuration points to interaction between the two sectors, and it implies active policy processes to fuel and shape that interaction. Third, the notion of configurations is broad enough to include more than just organizational arrangements. For example, configurations of information, including information about options, costs, and benefits, can provide foundations for arrangements that link the sectors.

In the context of research in the fields of evaluation and public administration more broadly, the notion of configurations also serves as one element in an emergent research agenda. That is, the study of privatization appears to be moving into a new phase of activity directed toward understanding how organizational arrangements affect outcomes. The sense of emerging direction is hardly new in policy science. For example, Goggin, Bowman, Lester, and O'Toole (1990) call for a "third generation" of research on policy implementation. Guba and Lincoln (1989) elaborate the foundations of a "fourth generation" of evaluation. The point here is simple: Scholars in specific subfields can find it useful to reflect on the stages of activity and conceptualization through which their craft has developed. The 1980s saw a burst of research activity in the field of privatization. Much of the work in this period focused on ideological arguments pro and con. Empirical research tended to focus on the premises of those arguments. Key issues had to do with economic efficiency and political accountability. Perhaps the one generalization clearly supported by all of the results is that in the field of privatization, generalizations are risky. Public-private partnerships, in their many forms, can provide efficiency and accountability, but they do not always do so. The conditions for success are *subtly dependent on context* (compare Guba and Lincoln's [1989] emphasis on precisely this issue).

Thus, the evaluation of privatization can productively address matters of context, including the matter of organizational arrangements. The chapters in this volume do so in a variety of ways. For example, O'Toole's (Chapter One) assessment of costs and performance under public and private management gives detailed attention to several factors, including extremely thorough and sensitive negotiation with the firm that provided him unusually rich data; the dependence of management outcomes on specific, idiographic, contextual details in individual cases; and the complexity and importance of management *strategy* in dealing with specific municipal settings. The simple message in O'Toole's findings is that privatization can work successfully, but it is difficult to generalize about the conditions under which this happens or can be made to happen. Perhaps the deeper message is that questions of a different kind are thus now in order.

The other three chapters offer such questions, and some answers as well. All deal in some way with options for configuring the dynamics that drive the public and private sectors (on these dynamics, see Heilman and Johnson, in press).

Spindler's (Chapter Two) contribution calls attention to legislative strategy as a mechanism for shaping private-sector activity in response to public-sector mandates. The contribution of this chapter is empirical as well as conceptual. Based on a broad review of evidence in the solid waste management field, and the details of solid waste legislation in Florida, Spindler illustrates how a legislative strategy can promote a form of public-private partnership. Conceptually, Spindler's framework for categorizing public and private initiatives as mandatory or voluntary suggests a whole series of ideas and questions about what "really happens" as public and private organizations interact (cell 5 of Table 2.3 in Chapter Two) to serve public ends.

Heilman and Johnson (Chapter Three) focus on the role of state regulatory agencies in the management of privatization. Of course, this role presents many issues; the question addressed in their chapter has to do with how state agencies approach the issue of privatization. The process that emerges is perhaps surprising because although it is diffuse, it turns out to have an identifiable structure. That is, individual state agencies do not have clearly defined game plans for addressing privatization. Their posture tends to be largely reactive. Nevertheless, there *is* a clear structure or sequence through which these organizations, in aggregate, proceed to address the subject. This finding may be of interest for two reasons. First, the trend toward increasing policy responsibility at the state level, and an increasing role for the private sector in the achievement of policy goals, seems likely to continue. Second, these findings indicate that information services, perhaps in the form of clearinghouses, can play a positive role in helping state agencies engage these matters.

Finally, Fitzgerald and McCabe (Chapter Four) illuminate the value of Weiss's (1988) conception of how evaluation is utilized (and, accordingly, done). In the management of nuclear waste, evaluation in the form of quality assurance (QA) plays a highly visible role, and high stakes are involved. Also involved are a complex assortment of organizational actors that both generate evaluation data and use the results in a politically controversial context. Fitzgerald and McCabe draw out the questions that this situation raises. What are the implications for evaluation when the process must be "perfect," the stakes are high, and the results will be controversial however they are used? What happens when the field of organizational interests to which Weiss directs our attention is excessively energized by conflicting values and perceptions?

The story of Yucca Mountain provides some partial answers that tie into themes raised in the Editor's Notes. First, that story makes clear that under intense political pressure and conflicting scientific expertise, normal conditions are suspended, or even turned upside down. In this context, Weiss's conception of utilization clarifies the dilemma of evaluation research: Such a broad issue-network of actors becomes involved in the "use" of results that a stalemate in the policy process can result.

The active entry of the private sector into the process also creates unexpected results. Part of DOE's problem in QA is that the scientific workload is so heavy that it has to be parceled out to numerous contractors, many or most in the private sector. The standard complaint against the role of the private sector in such circumstances is that it reduces public accountability. In this case, however, just the reverse appears to have taken place. That is, private contractors, by staking out independent positions and providing contradictory interpretations of the waste management potential of the Yucca Mountain site, increased public awareness of issues, and with it the level of accountability present in the process. In this process, as Fitzgerald and McCabe suggest, the evaluation contractors defined the terms of public debate over site acceptability.

Another interpretation of this outcome is that it engages the notion of reconfiguration of organizational arrangements, as discussed in the Editor's Notes. It is noteworthy that standard organizational steps within the public agency, such as those that DOE took to restructure the position of the QA function, appear insufficient in relation to the political demands of the situation. The entry of diverse private research firms into the agency-administered evaluative process, however, ensured that different perspectives could be developed and represented.

Constitutional issues are at stake. In referring to the Yucca Mountain controversy as possibly "the most intense battle over states' rights since the Civil War," Fitzgerald and McCabe indicate that we have here a constitutional application of evaluative research. It is no overstatement to suggest that in this case the foundations of democracy are being exposed. Their chapter clarifies the evaluation dilemma; responses to it remain to be worked out.

A few proposals that merit examination in this regard are already on the table. As suggested in the Editor's Notes, Wise (1990) and Kash (1989) urge attention to the careful configuration of public and private actors. A more limited and specific suggestion appears in the argument by Cronbach and others (1980) that there are benefits to reliance on authoritative assessment of evaluation work by experts in the field. The translation of these conceptions into substantive policy proposals remains to be undertaken. The role of the chapters in the present volume is to point to a range of challenges—and needs for well-designed strategies—that await policymakers and evaluators as the public and private sectors engage each other in the field of public service.

John G. Heilman
Editor

References

Cronbach, L. J., Ambron, S. R., Dornbusch, S. M., Hess, R. D., Hornik, R. C., Phillips, D. C., Walker, D. F., and Weiner, S. S. *Toward Reform of Program Evaluation: Aims, Methods, and Institutional Arrangements.* San Francisco: Jossey-Bass, 1980.

Goggin, M. L., Bowman, A. O., Lester, J. P., and O'Toole, L. J., Jr. *Implementation Theory and Practice: Toward a Third Generation.* Glenview, Ill.: Scott, Foresman, 1990.

Guba, E., and Lincoln, Y. *Fourth-Generation Evaluation.* Newbury Park, Calif.: Sage, 1989.

Heilman, J. G., and Johnson, G. W. *The Politics and Economics of Privatization.* Tuscaloosa: University of Alabama Press, in press.

Kash, D. E. *Perpetual Innovation: The New World of Competition.* New York: Basic Books, 1989.

Weiss, C. H. "Evaluation for Decisions: Is Anybody There? Does Anybody Care?" *Evaluation Practice,* 1988, 9 (1), 5–19.

Wise, C. R. "Public Service Configurations: Public Organization Design in the Post-Privatization Era." *Public Administration Review,* 1990, 50 (2), 141–155.

INDEX

Adams, S., 76, 86
Aluminum Company of America, 39
Ambron, S. R., 92
American Federation of State, County, and Municipal Employees (AFSCME), 14, 17, 30
American Iron & Steel Institute, 39
American National Standards Institute/ American Society for Quality Control (ASQC), 73, 86
Amoco Chemical, 40
Anheuser-Busch Companies, 39
Arco Chemical, 40
Armington, R. Q., 14, 30
Atomic Energy Commission (AEC), 70

Babbie, E., 62, 64, 66
Babitsky, T. T., 14, 31
Barke, R., 75, 86
Benveniste, G., 79, 86
Beverage industry: and deposit legislation, 41-44; and nonreturnable bottles, 35-36; recycling efforts by, 38-39
Beverage Industry Recycling Program (BIRP), 38
Borcherding, T. E., 14, 30
Bottle bills, history of, 35-36
Bowman, A. O., 89, 92
Brooks, H., 14, 31
Brudney, J. L., 14, 31
Bryan, R., 82
Byner, G. C., 84, 86

Calem, M., 43, 44, 46, 47
Can Manufacturers Institute, 39
Carey, R., 2, 10
Carroll, J. D., 54, 66
Carter, L. J., 75, 86
Chevron Chemical, 40
Claudson, T. T., 73, 86
Clean Water Act of 1972, 50, 51
Clinch River Task Force, 70
Colglazier, E. W., 75, 86
Collins, F. C., Jr., 73, 86
Compelled partnership, 36, 39-40; Florida example of, 9, 40-44

Contracting, 1, 14-15, 30; research on, 16-30; for wastewater treatment operations, 15-16
CRInc, 38
Cronbach, L. J., 92

Davis, J. A., 76, 86
Davis, T., 38, 40, 42, 47
DeHoog, R. H., 14, 31
DeMarco, J. J., 14, 31
Dickson, D., 75, 80, 86
Dietz, T. M., 81, 86
DOE. See U.S. Department of Energy (DOE)
Dornbusch, S. M., 92
Dorsey, T. A., 54, 66
Dow Chemical, 40

Edison Electric Institute/Utility Nuclear Waste Management Group (EEI/ UNWMG), 74
Ellis, W., 14, 30
England, J., 13n
EPA. See U.S. Environmental Protection Agency (EPA)
Escheats, 43
Evaluation: issues in, of privatization, 2-8, 89-90. See also Quality assurance (QA)

Farie, J., 14, 31
Feller, I., 54, 66
Fina Oil & Chemical, 40
Financial incentives: design of system for, 45-46; to manage solid waste, 41-44
Fitzgerald, M. R., 9, 14, 31, 69, 71, 77, 86, 87, 91-92
Florestano, P. S., 14, 31
Florida, compelled partnership strategy in, 40-44
Florida Soft Drink Association, 34, 41
Florida Solid Waste Act, 9, 41
Flynn, R. J., 54, 66
Folz, D. H., 77, 86
Franklin Institute, 46

93

ORDERING INFORMATION

NEW DIRECTIONS FOR PROGRAM EVALUATION is a series of paperback books that presents the latest techniques and procedures for conducting useful evaluation studies of all types of programs. Books in the series are published quarterly in Fall, Winter, Spring, and Summer and are available for purchase by subscription as well as by single copy.

SUBSCRIPTIONS for 1991 cost $48.00 for individuals (a savings of 20 percent over single-copy prices) and $70.00 for institutions, agencies, and libraries. Please do not send institutional checks for personal subscriptions. Standing orders are accepted.

SINGLE COPIES cost $15.95 when payment accompanies order. (California, New Jersey, New York, and Washington, D.C., residents please include appropriate sales tax.) Billed orders will be charged postage and handling.

DISCOUNTS FOR QUANTITY ORDERS are available. Please write to the address below for information.

ALL ORDERS must include either the name of an individual or an official purchase order number. Please submit your order as follows:
Subscriptions: specify series and year subscription is to begin
Single copies: include individual title code (such as PE1)

MAIL ALL ORDERS TO:
Jossey-Bass Inc., Publishers
350 Sansome Street
San Francisco, California 94104

FOR SALES OUTSIDE OF THE UNITED STATES CONTACT:
Maxwell Macmillan International Publishing Group
866 Third Avenue
New York, New York 10022

New Directions for Program Evaluation Series
Nick L. Smith, *Editor-in-Chief*